Environmental Science

Series editors: R. Allan · U. Förstner · W. Salomons

Springer

Berlin
Heidelberg
New York
Barcelona
Budapest
Hong Kong
London
Milan
Paris
Santa Clara
Singapore
Tokyo

Luiz D. de Lacerda · Wim Salomons

Mercury from Gold and Silver Mining: A Chemical Time Bomb?

With 44 Figures and 29 Tables

Springer

Dr. Luiz D. de Lacerda
Universidade Federal Fluminense
Dept. Geoquimica
Centro Niteroi
24020-007 Rio de Janeiro
Brazil

Professor Dr. Wim Salomons
GKSS Research Center
Max-Planck-Strasse
21502 Geesthacht
Germany

ISBN 3-540-61724-8 Springer-Verlag Berlin Heidelberg New York

Library of Congress Cataloging-in-Publication Data

Lacerda, Luiz Drude de.
Mercury from gold and silver mining : a chemical time bomb? / Luiz Drude de Lacerda, Wim Salomons.
p. cm. – (Environmental science)
Includes bibliographical references and index.
ISBN 3-540-61724-8 (hardcover)
1. Mercury – Environmental aspects. 2. Gold mines and mining – Environmental aspects. 3. Silver mines and mining–Environmental aspects. 4. Mercury wastes – Environmental aspects – Tropics. I. Salomons, W. (Willem), 1945– . II. Title. III. Series : Environmental science (Berlin, Germany)
TD196.M38L34 1998
363.738′4–dc21 97–35099

Typesetting: Fotosatz-Service Köhler OHG, Würzburg
Cover layout: Struve & Partner, Heidelberg

SPIN: 10551435 32/3020 – 5 4 3 2 1 0 – Printed of acid-free paper

Preface

Mercury contamination is considered one of the worst hazards among the anthropogenic impacts upon the environment. It is one of the few metal pollutants which has already caused human deaths due to ingestion of contaminated food. It is estimated that in the whole world more than 1400 human beings have died and over 20 000 have been afflicted due to mercury poisoning, with mortality rates ranging from 7 to 11% (WHO 1976; D'Itri and D'Itri 1977; Lodenius 1985; D'Itri 1992).

Mercury is easily transformed into stable and highly toxic methylmercury by numerous microorganisms. This Hg species typically shows very long residence times in aquatic biota, resulting in severe contamination of fish in many regions. This can lead to serious economic problems for populations depending on fisheries and other aquatic resources. It was in the fishing village of Minamata, in Japan, where between 1956 and 1960 more than 150 people died and over 1000 were handicapped for life due to consumption of mercury-contaminated fish. The contamination started by the release of relatively harmless mercury compounds. However, under the anaerobic conditions prevailing in the local aquatic sediments, bacterial activity transformed inorganic Hg into the highly toxic methyl-Hg compound (Fujika 1963). By December 1987 more than 17000 persons had been affected by methyl-Hg poisoning and 999 individuals had died (D'Itri 1992). Later, the same situation caused 6 deaths and 47 seriously ill in 1964 in Niigata, Japan. For the first time, the magnitude and seriousness of the problem of mercury contamination were understood. It is noteworthy that Hg inputs in Minamata Bay started some 40 to 50 years before the first symptoms of Hg poisoning appeared in the population. Other examples of severe Hg poisoning occurred in Iraq in 1972, where wheat seeds coated with methyl-Hg salts for protection against fungi were consumed by the rural population in the provinces of Ninevah, Kirkuk and Acbil. Over 6000 people were poisoned and 459 died (Bakir et al. 1973). In Brazil in the late 1970s, Hg also caused serious health problems among sugarcane farmers who used Hg compounds for sugarcane seedling protection (Camara 1985, 1986).

Although on a smaller scale, contaminated seeds used to feed pigs in New Mexico, USA, also caused serious health problems in farmers who had consumed pork with high concentrations of methyl-Hg (Mitra 1986).

This book focuses on the utilization of mercury in gold and silver mining. Once widespread worldwide up to the beginning of the present century, it faced a decline to near cessation due to the exhaustion of gold- and silver-rich reserves in the Americas and later to the invention of cyanide leaching. Recently, however, a confluence of economic and social situations, mostly in developing countries located in the tropics, has resulted in a new rush for gold and silver by individual entrepreneurs, for whom mercury amalgamation is a cheap, reliable and easy to carry out operation. This use of mercury is associated with large losses to the environment, in particular to the atmosphere. In this way, it ends up in places far away from where it is in use. Moreover, where gold mining occurs, hot spots of mercury remain. Both the hot spots and its widespread distribution in the environment cause effects which have been called colloquially chemical time bombs. The chemical time bomb concept deals with the limited capacity of soils and sediments to reduce the mobility and bio-availability of pollutants. Several properties (capacity-controlling parameters) of the soil, e.g. organic matter content and variables like pH and redox, contribute to these inherent soil and sediment properties. Although these capacity-controlling parameters can be reduced, even with no additional contaminant loading, negative effects on the environment may occur (Salomons and Stigiliani 1995). For example, in Scandinavia, mercury levels in freshwater fish are increasing even though emissions and deposition of mercury have been decreasing for several decades. The contamination of fish has been attributed to the remobilization of mercury locked in watershed soils decades ago, and now mobilized by soil acidification.

Although much information is already available on the fate and effects of mercury in temperate climate systems and on delayed effects (Salomons and Stigliani 1995), relatively little information is available for the often more sensitive tropical ecosystems. In this book, the widespread use of mercury in gold mining, its distribution in the tropical environment, and its impact on the ecosystem and on humans are documented for the first time.

The decision to write this book was made after a visit by L. D. Lacerda to Haren, Holland, to work with W. Salomons on a report on the situation of the Amazon region regarding mercury contamination. The visit, in the summer of 1991, was sponsored by the Dutch Ministry of Housing, Physical Planning and the Environment. In the following years, it became clear that mercury contamination due to gold mining was a global rather than a local phenomenon. At various scientific meetings since 1991, we have had the opportunity of talking to scientists from many countries where the problem was just beginning, and the resultant picture was quite alarming. This book, therefore, is not intended to be a definitive work on the subject, but tries to review in an integrated way the present knowledge on mercury contamination due to gold mining.

Both of us worked together in the Amazon region, and this was fundamental to the development of many ideas expressed here. This fieldwork was supported by many Brazilian institutions, in particular the Conselho Nacional de Desenvolvimento Científico e Tecnológico (CNPq, Brazil), the Centro de Tecnologia Mineral (CETEM, Rio de Janeiro) and the Universidade Federal Fluminense (UFF, Niterói). The possibility of working together in the field was unique for the development of this book, and the Brazilian and Dutch governmental agencies are gratefully acknowledged for their support.

Many colleagues throughout the world provided us with original information, original data and manuscripts. Among them special thanks are due to R.V. Marins, S. Rodrigues and R. Melamedi (CETEM, Rio de Janeiro); W.C. Pfeiffer, O. Malm and J.R.D. Guimarães (Inst. Biophysics, Rio de Janeiro); P. Lechler (University of Nevada, Reno); C. Ming (China); Y. Ykingura (Tanzania).

We are particularly grateful to R.V. Marins for carefully reading and commenting on early versions of this book.

Luiz D. de Lacerda
Wim Salomons

Contents

1 The Use of Mercury Amalgamation in Gold and Silver Mining

1.1 Historical Background

Metallic mercury has been known to man from at least 3500 B.P. The famous archaeologist, H. Schliemamm, discovered a small vessel full of mercury in a grave at Kurna, Egypt, dating back to 1600 to 1700 years B.C. Cinnabar, the primary mercury-bearing ore composed of mercuric sulfide, has probably been used as pigment since prehistoric times. The use of Hg in the mining industry to amalgamate and concentrate precious metals probably dates back to the Phoenicians and Carthaginians, who commercialized Hg from Almadén mines in Spain as earlier as 2700 B.P. Pliny, in his *Natural History*, provided the first detailed description of the amalgamation process as a common gold and silver mining technique at the beginning of the present era. This technology, however, had widespread use only by the Romans around the year 50 A.D. (Mellor 1952). Analyses of these descriptions, dating back nearly 2100 years, show distinct similarities with the procedures presently applied in many gold mining areas in the world.

Environmental problems caused by the utilization of Hg in gold and silver mining have been reported since Roman time. Around 77 A.D. Roman authorities were importing circa 5000 kg/year of mercury from Spain to be used in gold amalgamation in Italy. Curiously, after less than 100 years, this activity was forbidden in mainland Italy (D'Itri and D'Itri 1977; Nriagu 1979). It is quite possible that this prohibition was already a response to environmental health problems caused by the activity.

Up to the sixteenth century, Hg amalgamation was the major technique used for the mining of precious metals. However, the small amounts of precious ore present in the known Western world prevented any significant release of Hg in the environment. Outside Europe, amalgamation may have been used by pre-Columbian Americans, but available evidence (high Hg concentrations in gold-plated artifacts from that period) is questionable

(Patterson 1971). After the Spanish invasion, however, the situation changed completely.

Prospecting of cinnabar deposits were strongly supported in the New World by the Spanish Colonial Government. With the development of the "Patio" amalgamation process by Bartolomeu de Medina in 1554 in Spanish Mexico, and its later introduction to silver mines in Mexico, Peru and Bolivia, mercury amalgamation reached its peak. The "Patio" process consists of spreading silver- and gold-powdered ore over large, paved, flat surfaces and mixing it with salt brine and a mixture of Cu and Fe pyrites and elemental mercury. Workmen or mules blend the mixture with hoes and rakes and let it stay for days to weeks, depending on the weather, for amalgamation. After removing the amalgam, Au, Ag and Hg are recuperated through roasting (Fig. 1.1).

The silver production in colonial America, in particular in Mexico and Peru, was an impressive conveyor belt of Hg to the environment. In 1870, over 70% of Mexican silver was produced through this technology. In the Potosi mines between 1545 and 1803, over 25000 tons of silver were produced through Hg amalgamation. The town, by the time the peak of the mining operation was reached, had over 6000 small furnaces to produce this metal, and Hg intoxication was common place (Galeano 1981; Brüseke 1993). Production of major Hg mines was dedicated to amalgamating silver ore in the New World. Figure 1.2 shows a resumé of Hg production from the

Fig. 1.1. "Patio" amalgamation in silver and gold extraction in colonial Spanish America. (Redrawn from Mellor 1952)

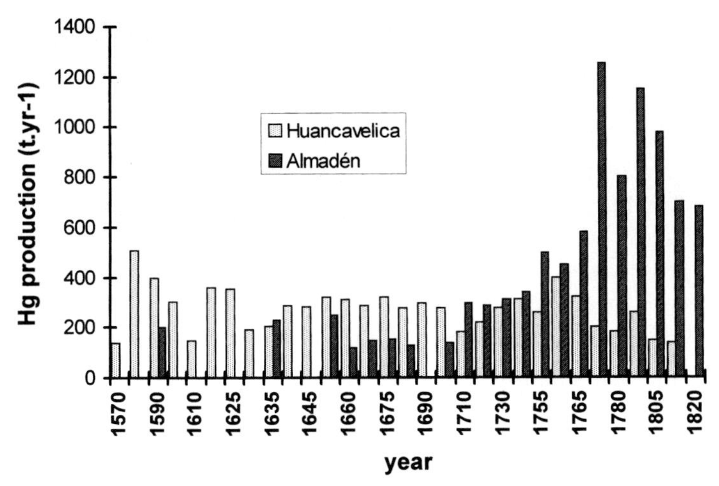

Fig. 1.2. Mercury production from the Spanish mines of Huancavelica (Peru) and Almadén (Spain) between 1690 to 1820. Data are average annual production for 5-year periods. (After Brading and Cross 1972)

Spanish mines of Huancavelica (Peru) and Almadén (Spain) used in silver mining in the New World. Between 1560 to 1700 over 20 000 tons of Hg were imported by Mexican mines, while Hg consumption in Peru reached over 50 000 tons in the same period. After 1700, while Hg production in the Peruvian Huancavelica mines stabilized at an average of 240 tons/year, Almadén increased to over 620 tons/year, resulting in an additional consumption of over 100 000 tons in the New World silver mines between 1700 to 1820 (Brading and Cross 1972). Although Spanish America dominated the world silver market, due to the use of Hg amalgamation, during the 1700s and 1800s, North America was a major gold producer using this technique to recover precious metals during the last century. Between 1850 and 1900, gold mining in the USA alone consumed from 268 to 2820 tons/year of Hg, totalling nearly 70 000 tons for that period. The gold rush to western US states was also responsible for the exportation of this technology to Australia, Canada and other countries. In fact, most mining equipment in use in gold mining today throughout the tropics was developed originally in the gold fields of North America (Nriagu 1994).

Once widespread worldwide up to the beginning of the present century, this "technology" faced a decline to near cessation due to the exhaustion of the rich silver and gold deposits in the Americas and the invention of cyanide leaching (Rose 1915; Mellor 1952; Nriagu 1993).

After the development of cyanidation in the 1880s and its global introduction by 1990 for mining gold from low-grade deposits, and the exhaustion of the silver mines in colonial America and of the rich alluvial gold deposits in the US, mercury amalgamation virtually disappeared as a significant mining technology up to the 1970s, when it was re-introduced in the tropics.

1.2 Present-Day Worldwide Utilization of Mercury Amalgamation

Recently, a confluence of economic and social situations, mostly in developing countries located in the tropics, has resulted in a new rush for gold and silver by individual entrepreneurs, for whom mercury amalgamation is a cheap, reliable and easy to carry out operation.

An impressive increase in gold prices, from US $58.16 oz t in 1972 to US $446.56 oz t in 1987, triggered a second gold rush in many tropical countries, in particular in Latin America and Asia, and more recently in Africa. In Brazil in particular, Hg amalgamation was responsible for the production of only 5.9 tons of gold in 1973. In 1988 this "technology" was responsible for the production of over 100 tons of gold, mostly from the Amazon region and central Brazil. Today, minor operations are found even in the industrialized south and southeast regions (DNPM 1989; Pestana et al. 1993).

This true gold rush rapidly spread through other Amazonian countries, in particular Venezuela, where many rivers of the northern portion of the Guyana Shield have been deforested by 35 000 miners since 1980 (Mendoza 1990). Also in Colombia, this second gold rush was impressive. In this country, one-third of the gold production in Latin America, approximately 75% of gold production prior to 1970, came from industrial mines. Later, gold production from small-scale and artisanal mining, most of them using Hg amalgamation, became responsible for over 94% of the total gold production (Prieto 1995). Other important, but less studied sites are located at the Pando River Department, Bolivia (Zapata 1994), French Guyana (Chris Wood 1994, pers. comm.), the Puyango River at the Peru-Ecuador border, and in the gold fields of Choco and Nariño Departments in Colombia (CIMELCO 1991; Priester 1992).

In the Philippines, where small-scale gold prospecting similar to that in the Amazon region is responsible for nearly 80% of the country's gold production (Torres 1994), Hg amalgamation is believed to have produced over 50 kg of gold daily since 1985, employing over 800 000 people, particularly Mindanao Island (Cramer 1990). Some 26 tons of Hg are released annually into rivers draining into the Ayusan River to Butuan Bay, a rich fishing area in Davao del Norte Province (Torres 1992).

Small-scale mines, employing thousands of prospectors, are also present in Indonesia. In this country, an ingenious pebble mill and amalgamation drum are used for gold production. These mills, locally called "tromol" mills, are made and carried by the hundreds to remote mountain areas. As much as 2% Hg is added to a drum of approximately 50 kg of broken ore. Hundreds and probably thousands of such appliances are used daily throughout the country, mobilizing large amounts of Hg (James 1994). Small-scale mining is also found in Siberia, Russia; Thailand and Tanzania (Taysayev 1991; Achmadi 1994; Ikingura 1994; Fig. 1.3). Small-scale mining operations using Hg amalgamation still occur in Canada, and Australia and "weekend panners" are still typical of many areas in the USA.

After the permission was given for individual enterprises in China, over 200 small mines have been opened in Dixing Province since 1992, increasing gold production to 10% per year (Yshuan and Liu 1994). However, at least one large mine has been using Hg for gold production since 1938, in the Jia pi Valley, in JiLin Province, using the same techniques used in California in the 1800s (Ming 1994). The same phenomenon occurred in Tanzania, where since 1991 over 150 000 people have been prospecting gold in the Lake Victoria gold fields (Ikingura 1994).

At the peak of these activities in 1989, over 6 million people were believed to be participating in gold prospecting in the Amazon region alone and probably over 20 million throughout the world. Mercury poisoning due to gold mining has been reported for at least three major areas, the Brazilian Amazon, the Dixing region in China, and the Philippines, and will be discussed in detail later.

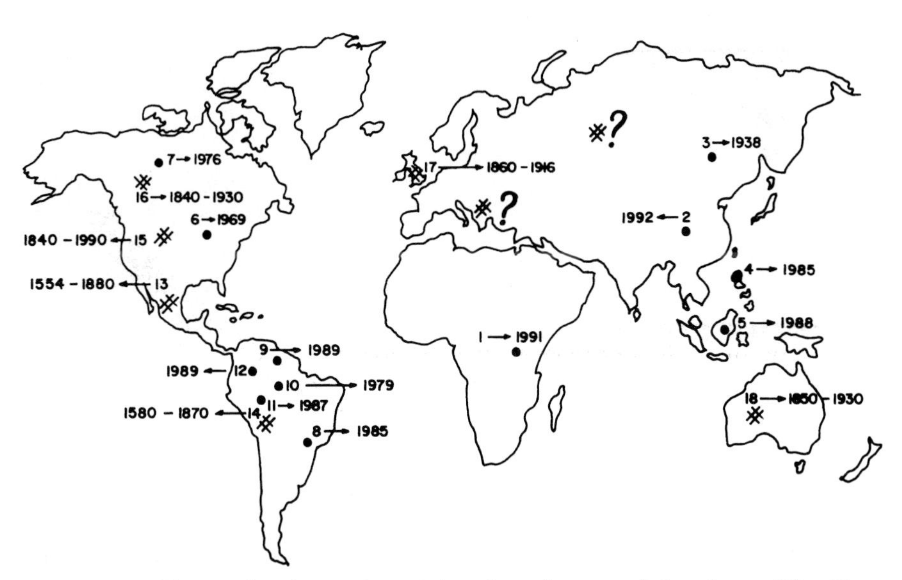

Fig. 1.3. World map showing major mining sites of present (●) and past (#) utilization of Hg amalgamation in gold and silver mining

1.3 Present-Day Amalgamation Mining Techniques

Mercury is used for the separation of fine gold particles through amalgamation after a gravimetric preconcentration step involving the heavy fraction of river sediments, soils or grounded rock, depending on the mining site. After the amalgamation step, the Au/Ag-Hg complexes are generally distilled in retorts, but in most areas this operation is carried out in the open air, thus emitting Hg to the atmosphere in large amounts. During the amalgamation process, a variable amount of metallic Hg is also lost to rivers and soils through handling under rough field conditions and due to vaporization. Also, Hg-rich tailings are left in most mining sites (Fig. 1.4 a, b). Details of procedures, instruments and processes presently in use in gold and silver mining using Hg amalgamation can be found in many publications (DNPM 1983; CETEM 1989; Garrido et al. 1989; Souza and Lins 1989; James 1994). All these have been nearly the same since the seventeenth century (see, for example, Rose 1915; Averill 1946; Mellor 1952; Wise 1966), and probably also similar to the techniques used in Roman times (Mellor 1952; Nriagu 1993 a,b).

The various procedures presently in use, however, can be grouped into two large categories.

Fig. 1.4. Typical tailings deposits at the Poconé mining site, Mato Grosso State. **a** Central Brazil; **b** Nariño Province, Colombia

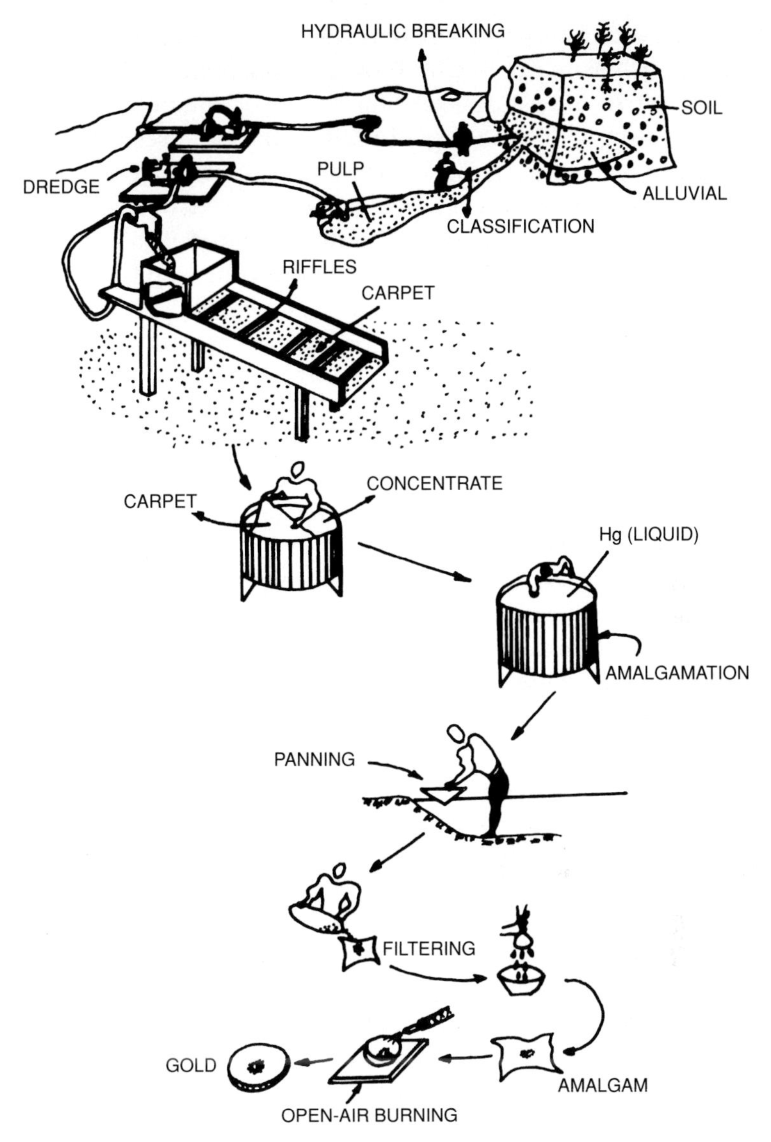

Fig. 1.5. Steps in gold and silver production and Hg utilization in soil and rock mining

The first involves the recuperation of gold and silver from soils and rocks, whereby the content of precious metals ranges from 4.0 to 20.0 g/ton. This procedure consists of digging up large amounts of metal-rich material, generally soils containing quartz veins or sulfides (Fig. 1.5), passing it through grounding mills and centrifuging to produce a metal-rich concentrate. In colonial America, mules and slaves were used instead of electric mills (Mellor 1952). This procedure may result in pronounced deforestation along rivers such as in the Nariño Province, Colombia and central Brazil, and may result in intense soil erosion and river siltation. In the Choco Department, also in Colombia, gold production increases 7.2% per year, resulting in deforestation rates of 1000 ha/year (CODECHOCO 1991). Suspended solids concentrations in Amazon rivers affected by this activity may increase up to 3 orders of magnitude (Rodrigues 1994).

The concentrate is moved to small amalgamation ponds, a few square meters large or to amalgamation drums, where it is mixed with liquid mercury and separated later in round pans. The Hg-Au/Ag amalgam is then squeezed to remove excess Hg and then taken to a retort for roasting (Fig. 1.5).

Alternatively, the powdered ore is floated with water as pulp in a thin stream over slightly inclined plates, generally copper plates, amalgamated with mercury. These amalgamation tables were first introduced in California in the 1850s and are still in use in many regions, such as in Colombia and northern Brazil (Priester 1993; Fig. 1.6).

Unfortunately, the use of retorts (Fig. 1.7b), although cheap and simple to use, is far from being a common practice in most mining regions. This is due to a low environmental awareness among the miners, disgard with respect to the toxicity of mercury itself and lack of legislation enforcement due to general, difficult logistics. Instead, the amalgam is more frequently roasted in pans in open air, thus releasing the entire Hg content of the amalgam directly to the atmosphere, and often resulting in accidents involving the inspiration of significant amounts of mercury vapor (Fig. 1.7a; Torres 1994).

Any residue of the concentrate is returned to the amalgamation pond and is reworked until the gold in the area is exhausted. Mercury is lost to the atmosphere during this process through vaporization of Hg drops left in pans exposed to the sun and from leaking of retorts. After an area is exhausted, tailings are left with spots of high Hg concentrations located in previous amalgamation ponds. In such spots, typical Hg concentrations can reach up to 30 to 80 µg/g of dry tailing material (Bycroft et al. 1982; CETEM 1989), but frequently range from 1 to 5 µg/g. Extreme values ranging from 500 to 4500 µg/g have been reported for old tailings in Canada and the USA respectively (Lane et al. 1988; Lechler 1993). In general, however, mean Hg concentrations in the majority of tailings material are very low, ranging from 0.04 to 0.2 µg/g (Andrade et al. 1988; Ramos and Costa 1990; Lacerda et al. 1991a,c), due to dilution with Hg-poor material such as quartz grains.

Soil upland alluvial mining involving the extraction of precious metals is widely used in central Brazil and northeastern Amazon, and southwest Colombia, where typical "moon-surface" landscapes and landslides are left behind as a result of mining (Fig. 1.8a, b). The lack of vegetation cover also

Fig. 1.6. Amalgamation tables in use in Nariño Province, Colombia

facilitates the degassing of Hg from these areas (Lechler 1993; Mason and Morel 1993, see also Sect. 3.4.1).

The second typical process is carried out in most Amazonian rivers (Fig. 1.9). Gold is extracted from bottom sediments by dredging. The gold-bearing material passes through iron nets of different mesh sizes to remove large stones. Grounding mills are not used in the operation. The material is then passed through carpeted riffles which retain heavier particles. This operation lasts for 20 to 30 h, then the dredging stops and the heavy fraction is collected in barrels for amalgamation, which can be done by hand or by using mechanical stirrers. The amalgam is then separated in the same way describe above. However, residues of the procedure are released into rivers. Vaporization of mercury and losses due to poor handling also occur (Lacerda et al. 1989; Fig. 1.10).

Regardless of the extraction process used, the resultant gold and silver still has a variable amount of mercury as impurity. The mercury concentration in roasted gold can reach up to 5% in weight. Therefore, wherever gold is sold, reburning (pyrolysis) is common practice. The process is performed when miners sell their gold to local gold dealers and without the required care of exhaustion and filtering of the contaminated air, resulting

a

b

Fig. 1.7. Roasting of gold-mercury amalgam in **a** open pans in the Madeira River mining site, Rondônia State, northwestern Amazon, Brazil and **b** the use of retorts (Pocone, Brazil)

in serious contamination of workplaces and significant atmospheric inputs of Hg in small towns where gold dealers concentrate (Malm et al. 1990; Marins et al. 1990, 1991). This represents a serious threat to local public health and since gold pyrolysis is actually a point source of Hg, it can be very important in determining Hg distribution in the natural ecosystems surrounding these towns (Hacon et al. 1995).

a

b

Fig. 1.8. "Moon landform" created as a result of the extraction of gold-containing soils in **a** Mato Grosso State, central Brazil and **b** in the Pantanal (Brazil)

a

b

c

Fig. 1.9a–c. Examples of alluvial gold mining. Dredging equipment in use at **a** the Madeira River, Rondônia State, northwestern Amazon, Brazil, **b** the Lo-An River in China and **c** the Kelian River in Indonesia

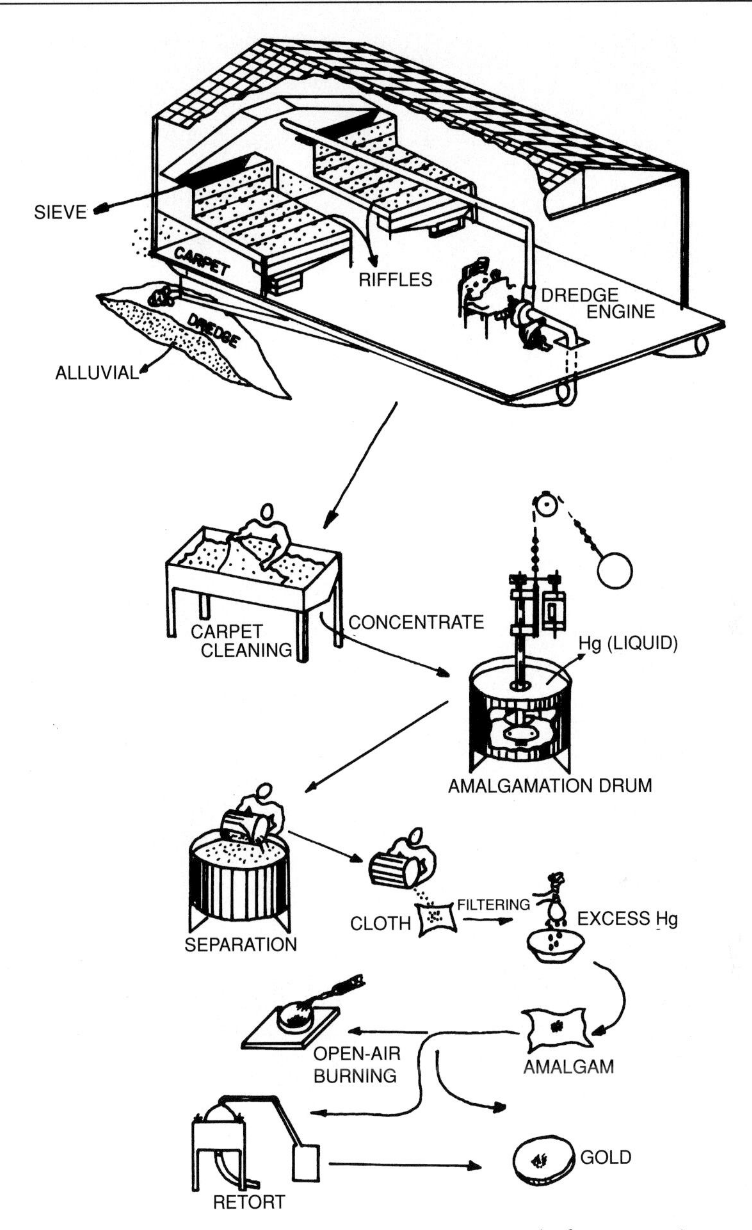

Fig. 1.10. Steps in gold production in river mining typical of Amazon rivers

2 Estimating Losses of Mercury to the Environment

2.1 Emissions Factors

The release of mercury from gold and silver mining is of great environmental significance, since this process of mining and production of precious metals is known to be of very low efficiency (Mallas and Benedicto 1986; Souza and Lins 1989; Nriagu 1993b). Correct estimates of the total amount of Hg released to the environment from these semi-artisanal mining processes, however, are very difficult to obtain. Gold and silver, on the other hand, are very precious metals and even in Roman times the quantities produced were carefully noted. Therefore, if the steps in the production and the Hg losses in each step are known, one could estimate more accurately the amount of Hg released into the environment using gold and silver production figures. *Emission factors* (EF), i.e. the amount of Hg released into the environment to produce 1.0 kg of silver or gold, are quite variable, depending on site, metal-containing material and concentration, and the extraction process used (Farid et al. 1991).

Estimates of emission factors have been published for many historical gold mining sites. Fisher (1977) estimated emission factors for the gold and silver production in Spanish mines in colonial South and Central America based on reports of mercury and gold and silver trade, as carefully noted by Spanish captains. Emission factors close to 1.5 were obtained. These EFs, however, could range from 0.85 for Ag-poor ores to 4.1 for very rich ones (Nriagu 1993b). Wise (1966) reported similar values for nineteenth century mines in Victoria gold fields in Australia and Mellor (1952) assumed emission factors higher than 2.0 for this technique.

Based on observation only, Cramer (1990) reported emission factors of nearly 5 in gold mining in the Philippines. CIMELCO (1991) reported emission factors of nearly 2.0 for Puyango River mining at the Peruvian-Ecuador border, similar to those reported by Priester (1993) for Nariño Province, Colombia, and along the Jia pi Valley in China (Ming 1994). However, the best studies available on this subject have been done through both observation and experimentation in the gold mining sites in the Brazilian Amazon.

The first EFs published on Amazon gold mining sites were based on observations by Mallas and Benedicto (1986) in Pará State, eastern Amazon, where gold is extracted from soil. These authors reported EF values ranging from 2.0 to 4.0 kg Hg/ 1.0 kg Au produced. However, these first estimates were based only on interviews with miners and field observations. Pfeiffer and Lacerda (1988), studying mining sites in the Madeira River, Rondonia, the largest tributary of the Amazon River, where gold is mined from active river sediments, found an average EF of 1.3, also based on interviews with several dredge crews, local geologists and empirical observations. In a large survey of 800 mining sites in central Brazil, The National Department for Mineral Production (DNPM) found an EF of ca. 1.7 kg Hg for each kg Au produced. These estimates were obtained by actually measuring the mercury balance throughout the whole process of gold production (Hacon 1991). Farid et al. (1991) measured mercury losses through the different steps of gold production in six mining sites in northern Mato Grosso State, central Brazil. They found EFs ranging from 0.1, where closed amalgamation systems and the use of retorts for roasting the amalgam recuperate most of the mercury used, to 1.1, where no such devices were in use, as in most previous reports. In the Mindanao region, the Philippines, Cramer (1990) reported that to produce nearly 15 kg of gold, 100 000 miners release nearly 26 000 kg of mercury into creeks and rivers of the region. These numbers would roughly reach an emission factor of circa 5 kg Hg/ kg gold produced (EF = ± 5).

Notwithstanding the variability of the reported EFs, they consistently fall between 1.0 to 1.7 kg Hg/ kg Au produced; also, all studies agree that Hg EFs to the atmosphere are much higher than those to rivers and soils, accounting for 65 to 83% of the total emission. These emission factors are the highest ever reported for any Hg-emitting activity (Nriagu and Pacyna 1988). For comparison, Hg EFs from chlor-alkali plants, a major source of Hg in the 1970s, was about 5–10 g Hg/ ton of chlorine produced (Pacyna and Münch 1991), with a maximum of 150 to 300 g/ton of chlorine in old plants (Mitra 1986; Bezerra 1990). For the production of other non-ferrous metals, such as Pb and Zn, typical EFs range from 3 to 45 g/ton of metal produced (Nriagu and Pacyna 1988; Pacyna and M¸nch 1991). All these are orders of magnitude lower than the EFs reported for gold and silver production through Hg amalgamation.

The emission of mercury to different environmental compartments during the whole extraction process has been studied by some authors. Pfeiffer and Lacerda (1988) estimated Hg losses to the environment in a mining site in the Madeira River, Rondônia State, NW Amazon. They estimated that the major proportion of Hg loss is during the burning of the Au-Hg amalgam, which in the area is frequently made without the use of retorts, amounting to 50 to 60 % of the total Hg loss, an extra 5 % is vaporized during the various extraction steps. Forty to 50 % Hg was lost to rivers during the amalgamation process as metallic Hg, and 5 to 10 % during the recuperation of the mercury used in the process also to rivers. In certain areas at this mining site, Hg is added directly into riffles, causing a significant increase in Hg loss to rivers as metallic Hg. These authors estimated the EF for this mining site to be circa 1.32 (Fig. 2.1).

The National Department for Mineral Production (DNPM), in its study of nearly 800 mining sites in central Brazil during 1987, where gold is mined from soils and rocks, found that nearly 87 % of mercury loss was to the atmosphere during the burning of the amalgam without the use of retorts and 13 % was lost to tailings (Pfeiffer and Lacerda 1988; Hacon 1991; Silva et al. 1991).

Farid et al. (1991) reported in a detailed study on Hg balance at six mining sites which use retorts for roasting the Au:Hg amalgam. They found that up to 70 % of Hg used in the process may be lost to the atmosphere if retorts are not used, while 20 % is lost as metallic Hg to tailings and 10 % is also lost to the atmosphere during the purification of the gold produced. The use of retorts at these sites, however, reduced atmospheric emissions from 70 % to values ranging from 1–49 %, depending on the retort used, with a mean reducing value of 20 % among all sites. It is

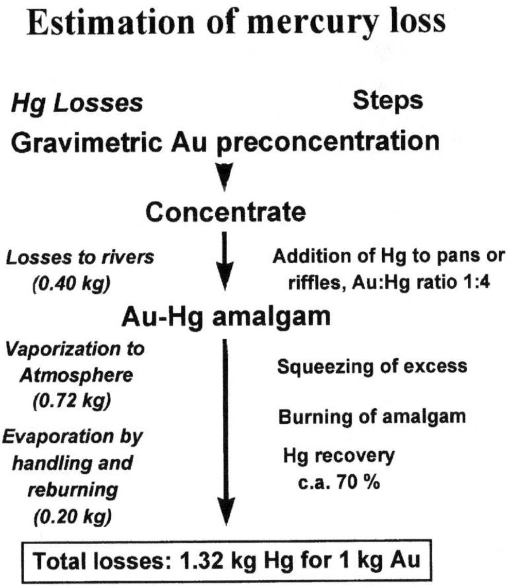

Fig. 2.1. Mercury losses to the environment during the various steps of gold extraction. Based on data from mining sites of the Madeira River, Rondônia State, Brazilian Amazon. (Pfeiffer and Lacerda 1988)

obvious that the use of retorts would significantly decrease Hg emissions to the atmosphere. Unfortunately, however, the use of such simple equipment at most gold mining sites is still very rare. Since 1988, due to society's pressure upon the government, gold dealers and miner's unions, an intensive campaign to introduce retorts and environmentally sound practices has been carried out by organized miner's unions. The success of such a program, which is restricted to areas were mining activities are more organized, however, is still unknown.

2.2 Mercury Emissions from Historical Gold and Silver Mining

It is possible to estimate Hg released to the environment during this period. First, one has to know the amount of Au and Ag produced, a carefully kept number by Spanish captains in colonial America for example, and second, the amount of Hg lost to the environment to produce that amount of Au or Ag. Estimates of Hg released to the environment in the production of precious metals is possible due to the utilization of Hg imported from only three sources (Amaden, Spain; Huancavelica, Peru; and Idria, Slovenia) and the similarity of the mining process used in colonial America to present-day procedures in use in many countries. Thus, emission factors can be calculated. Fisher (1977) calculated emission factors for Spanish silver and gold production using the Patio process with circa 1.0 kg Hg kg Au^{-1} or Ag^{-1}. The quantities of gold and silver produced were carefully noted by Spanish mine captains. Thus fairly reliable estimates can be derived. These estimates of Hg utilization and the consequent release into the environment during the extraction of silver and gold in central and northern South America during colonial times were calculated by Nriagu (1993a, b) and reached 196000 tons of Hg released into the environment between 1570 and 1900, with annual inputs ranging from 292–1085 tons (Nriagu 1993b; Fig. 2.2).

This high input of 260000 tons (ca. 80 t Hg $year^{-1}$) must have contributed to an increase in the background concentrations of mercury on this continent (Nriagu 1993b; Bloom and Porcella 1994), since from 180 to 705 tons/year of Hg were directly discharged into the atmosphere, leading to a regional and possibly global distribution (Nriagu 1993b).

The late 1800s was the time during which North America and Australia saw its gold rush and the Hg problems associated with it (Fig. 2.3). Mercury was also used predomantly in the 1850s in California, Nevada and South Dakota and later in Canada.

Approximately 90% of all Hg consumed in the US between 1850 and 1900 (1360 tons $year^{-1}$) was used in gold and silver mining. Over 61000 tons of Hg were emitted to the environment during that period (Nriagu 1994).

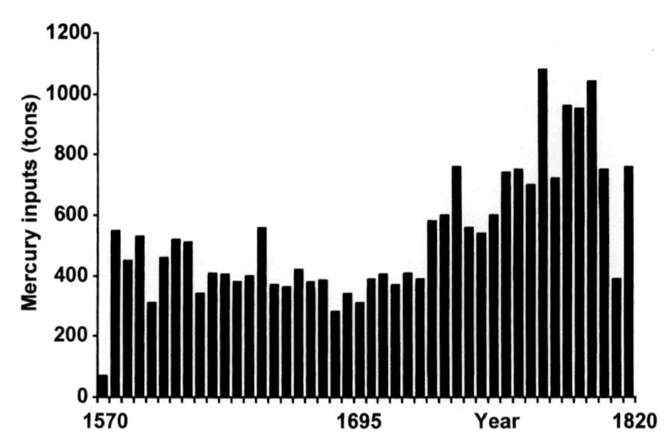

Fig. 2.2. Estimated annual inputs of Hg from silver production in colonial America. (After Nriagu 1993b)

Dredge tailings from the last century still cover over 73 km^2 in the Folsom-Natomas region in California. Since part of it is already urbanized, Hg contamination is a potential threat to people living there (Averill 1946; Prokopovich 1984). The contamination of soils and waters by mercury, as well as poisoning of miners, in particular those in charge of roasting Au-Hg amalgam, has been reported in both regions. Tailings, produced as early as 1840 and containing Hg of up to 500 $\mu g\ g^{-1}$ of Hg, are still present in Nova Scotia, Canada (Lane et al. 1978), while in the Carson River area over 100000 tons of tailings containing Hg with concentrations of up to 4900 $\mu g\ g^{-1}$ are being continuously reworked along the Carson River Valley,

Fig. 2.3. Pictorial representation of the California gold rush, showing major mining techniques. (Redrawn from Mellor 1952)

potentially mobilizing over 7000 tons of Hg from mining tailings (Lechler and Miller 1993).

In the South Mountains gold fields, North Carolina, fluvial sediments downstream of an old Au refinery built in the 1830s, and closed 30 years later, still present anomalously high, up to 7.4 $\mu g\ g^{-1}$ Hg concentrations, even after more than 100 years after the closing of the refinery. Exceptionally high Hg concentrations (up to 4.9 $\mu g\ g^{-1}$) in moss samples collected in that area suggest that the old tailings in that drainage area are still a present and significant source of Hg to the local environment (Callaham et al. 1994). Recently, Nriagu (1994) estimated that the total Hg load from the American gold rush during the 1800s was nearly 60 000 tons.

In Australia, alluvial gold mining using mercury amalgamation started in Victoria Territory by 1850 and lasted until 1930 (Smith 1869). Significant amounts of Hg are still contained in these old gold fields. Tailings from abandoned gold mines on the Thonson River, Victoria, contained from 40 to 90 $\mu g\ g^{-1}$ of Hg (Baycroft et al. 1982). At the Bendigo field, one of the largest in Victoria, up to 900 tons of Hg are estimated to have been lost to the environment between 1850 and 1930, while in the Lerderderg River, nearly 5 tons $year^{-1}$ of Hg were lost during this period (Baycroft et al. 1982). Small operations also took place in the United Kingdom. From 1860 to 1916, 3.7 tons of gold were produced through Hg amalgamation in the Dolgelau gold belt, North Wales, leaving behind tailings containing up to 6 $\mu g\ g^{-1}$ of Hg. These tailings are still a significant source of Hg contamination to local rivers (Fuge et al. 1992). Some Dutch mines in Indonesia have been using Hg amalgamation since the late 1800s, either in Hg-coated copper plates or added during the crushing of the ore (Lock 1901).

In Brazil, another major gold producer during the seventeenth and eighteenth centuries, mercury has a secular use in gold mining. Gold was first discovered in 1552 but its commercial exploitation started only in 1700 with the discovery of rich fields in central Brazil. The exploitation of these large deposits did not need a concentration step and lasted 100 years. By 1800 Brazil had produced over 830 tons of gold, corresponding to roughly 60 % of the global gold production at that time. After the exhaustion of this first gold cycle, Hg was introduced to mine low-grade ore deposits, but gold production was no longer significant and up to the late 1960s typical gold production in Brazil was always below 5 tons $year^{-1}$, with less than half of it produced using Hg amalgamation (Simonsen 1962; Hanai 1993). Therefore, although present, Hg inputs during the last 200 years in Brazil may have reached less than 500 tons, an extremely low amount when compared to the nearly 200 000 tons believed to have been lost in colonial Spanish America.

After this first gold rush, which ended by the earlier 1900s, Hg was still used in North America at a rate of 8 tons $year^{-1}$ in gold and silver mining up to the 1950s and from 4 to 6 tons year^{-1} during the 1960s and early 1970s (USGS 1968, 1970). Low, but continuous, utilization also occurred in Nova Scotia, Canada (Lane et al. 1978; Moore and Sutherland 1980). From 1968 to 1976 approximately 4.1 tons of Hg were still used for gold amalgamation, with yearly consumption ranging from 0.19–0.80 tons (Hocking 1979). An Environmental Protection Agency sampling of the Housteke Mining Company, South Dakota, USA, carried out in 1970, showed that this company, which has been operating since 1880, was still discharging from 5.5 to 18.0 kg of mercury daily into the Whitewood Creek and Cheyenne River system (EPA 1971), causing serious contamination in sediments and biota of these rivers (Walter et al. 1973). Through gold mining in North Carolina, "weekend panners" are still a significant source of Hg (Callaham et al. 1994).

Part of the total load of Hg to the environment in the last four centuries (Table 2.1) has ended up in the ocean. Since the oceanic residence time of Hg is approximately 350 years, much of the Hg emitted over the past 400 years still resides in the ocean and is potentially available for methylation, biological uptake, remobilization and cycling to the atmosphere. This input may have increased the background concentrations of Hg even at a global scale (Bloon and Porcella 1994).

Table 2.1. Estimates of past mercury annual (ton year^{-1}) and total emissions to the environment through gold and silver production

Site	Period	Annual input	Total input	Author
Roman Empire	77 AD	?	?	D'Itri and D'Itri (1977)
Cajamarca, Colombia	1948–1977	?	?	Alvarez-León (1994)
Spanish colonial America	1554–1880	292–1085	196000	Nriagu (1993 a,b)
Carson River, Nevada, USA	1850–1900	200	7000	Lechler (1993)
California, USA	1840–1900	17	1000	D'Itri and D'Itri (1977)
Nevada/California, USA	1900–1950	8	400	USGS (1968)
South Dakota, USA	1880–1970	2–7	380	Walter et al. (1973)
All North America	1840–1900	1000	60000	Nriagu (1994)
Bendigo Fields, Australia	1850–1930	11	900	Baycroft et al. (1982)
Colonial Brazil	1800–1960	5	400	Lacerda (1995)
Gwynedd, Wales, UK	1860–1916	0.1	6	Fuge et al., (1992)
Lerderderg River, Australia	1855–1930	5	400	Baycroft et al. (1982)

2.3 The Second Gold Rush in the South

After the development of the cyanide leaching process, Hg amalgamation virtually disappeared as a significant mining technology until the 1970s. A dramatic increase in gold prices, from US $58.16/oz t in 1972 to US $436.6/oz t in 1985, triggered a second gold rush in many tropical countries, particularly in Latin America. In Brazil, Hg amalgamation was responsible for the production of only 5.9 tons of gold in 1972. In 1989, this technology was responsible for the production of nearly 100 tons of gold, mostly from the Amazon region. This new gold rush spread rapidly through other Amazonian countries (in particular Venezuela, Peru and Colombia), and later to the Philippines, Thailand and Tanzania.

Brazil presents the best-studied case of this new gold rush and its effects using Hg amalgamation to extract gold. The development of this process is, in general, similar to many other tropical countries where this technique is being employed. Therefore, a description of the Brazilian case will be of great help in understanding future developments and environmental impacts of this new gold rush on a global scale.

In Brazil, the earlier "gold cycles", from 1690 to 1850, used only gravimetric separation procedures to extract large gold particles. Mercury is known to be used only in recent "gold cycles", which started in the 1960s following the increase in the international gold prices, making it economically feasible to prospect sediments and soils having a low gold concentration (Hacon et al. 1990; Pinto 1990; Veiga and Fernandes 1991).

The first "gold cycle" in Brazil occurred in the southeastern region using slave work to extract gold from ores with high concentrations. Nowadays, in these areas, industrial, large-scale gold mining occurs, although presently of less importance to the total Brazilian gold production. In the last decade most gold production was carried out by non-organized prospectors ("Garimpos"), activeat nearly 2000 sites spread throughout the Amazon region, where potential gold reserves are estimated to be at least 25000 tons, equivalent to nearly US $300 billion (Veiga and Fernandes 1991).

The recent surge in Amazonian gold mining has placed Brazil second worldwide in gold production (238 tons in 1988, of which 216 tons, over 90%, were produced by "garimpos" in the Amazon); behind South Africa (621 tons) and ahead of the USA (205 tons), Australia (152 tons) and Canada (116 tons). These five countries contribute over 85% of the global gold production of Western countries (Fernandes and Portella 1991).

The socio-economic significance of the Amazonian "garimpos" can be easily seen when compared with other economic activities in the Amazon.

In 1988, 216 tons of gold were produced by Amazonian "garimpos", equivalent to nearly US $3000 million, which corresponded to 18 % of the total Amazon GNP, or three times as valuable as the iron production of the Carajás mining site, one of the largest in the world. Also in 1988, 600 000 people worked in "garimpos", compared to a total of 2.1 million jobs available in the entire Amazon. This is also six times higher than the total amount of jobs available in the whole Brazilian mining industry (Fernandes and Portela 1991). Preliminary data for the year 1990 show no modification of this picture, with gold production of over 100 tons and circa 1 million people involved in Hg-based gold mining (Pinto 1990). The use of mercury is the major factor responsible for this accelerated growth in gold production in Brazil.

Brazil does not produce any mercury, thus the total amount of mercury used in the country is imported. The import of mercury was nearly constant from 1972 up until 1984, when it started to increase continuously, reaching a volume of 50 % more than the volume in 1986. Up to 1984, Mexico was the major Hg-exporting country to Brazil, reaching nearly 84 % of total Hg imported; the other 26 % was shared by the USA, Canada, and a few European countries. Presently, most mercury is imported from European countries, in particular The Netherlands, Germany and England, reaching nearly 340 tons in 1989. It is interesting to note that the shift from mercury-producing to non-producing European countries was coincident with the peak in the gold mining rush in the Amazon. For amalgamation purposes Hg is sold in relatively small quantities to a great number of individual miners. Most of the Hg imported by Brazil since 1984 has been for re-sale and non-specified commercial uses. This seems to explain the shift to countries selling Hg to Brazil (Ferreira and Appel 1991; Lacerda and Salomons 1991).

Figure 2.4 shows the evolution in the use of Hg in Brazil, based on declared uses during importation, since Brazil produces no mercury itself. It is clear that the use of Hg in chlor-alkali plants has decreased steadily over the last decade, from nearly 50 % of the total Hg consumed in Brazil in 1979 to less than 5 % in 1989, which is presently insignificant with regard to the total Brazilian mercury import statistics. The development and diversification of the Brazilian industrial park resulted in increases in Hg import for various uses, doubling the total amount used in the country. However, the total import of mercury for re-sale and for non- specified uses of 24 tons in 1979 increased to 78 tons in 1989, a 3.3-fold increase compared to only 1.9-fold for all other uses. It is believed that most of this mercury ends up in Amazon gold mining.

The exact calculation of the annual and total Hg lost to the environment, however, is still hampered by the lack of confidence in the estimates of

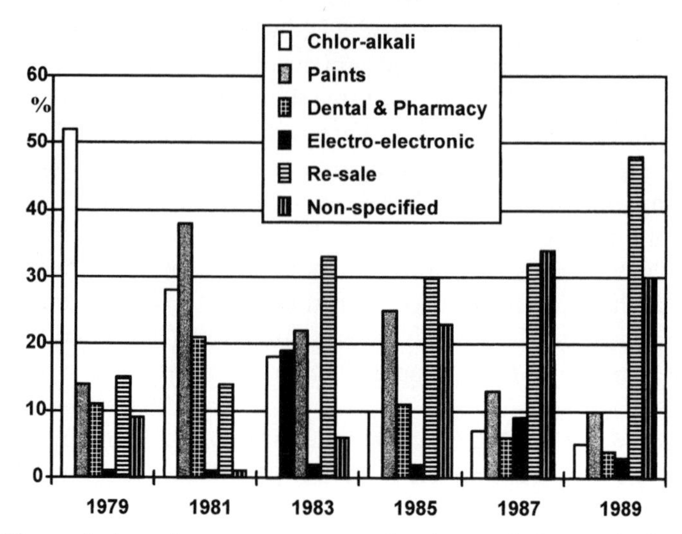

Fig. 2.4. The evolution of mercury consumption in Brazil during the last decade. Data are in percent of the total consumption for each year. (Ferreira and Appel 1990, 1991; Laborão 1990)

Brazilian gold and silver production in most mining regions. In Brazil for example, although there is dispute among official gold dealers, miner co-operatives and the Brazilian Department of Mineral Production, they all agree that from 40 to 100 tons of gold have been produced annually in the Amazon through mercury amalgamation techniques (DNPM 1983; Pfeiffer and Lacerda 1988; CETEM 1989; Lacerda et al. 1989; Souza and Lins 1989 Hacon 1991). Based on these published data and on personal observations in many gold mining sites in the Amazon, tentative estimates of gold production, Hg emission factors and Hg inputs into the Amazon environment could be made (Table 2.2).

These estimates agree with the data of Ferreira and Appel (1990, 1991) on the amount of Hg used in Brazilian "garimpos", and with the different figures presented by previous authors and also with mercury importation statistics for that period in Brazil. Therefore, notwithstanding the intrinsic variabilities of such calculations, these numbers seem to be fairly reliable and may be used for estimating Hg emissions from other mining areas in operation in other regions in the tropics.

Keeping in mind the uncertainties of such estimates, the total amount of mercury released annually in the Amazon environment (ca. 180 tons), for example, is four times the total anthropogenic Hg input for the highly industrialized North Sea basin (Salomons and Förstner 1984), and in the same order of magnitude of the annual Hg input to Europe as a whole (Pacyna and Münch 1991). In relation to atmospheric emissions only,

Table 2.2. Annual gold production, Hg emission factors and total Hg loss to the atmosphere and to rivers and soil in the Amazon region. Data are based on estimates of Pfeiffer and Lacerda (1988); Lacerda et al. (1989); DNPM (1983, 1988) Silva et al. (1991); Hacon (1991); Souza and Lins (1989); Ferreira and Appel (1990, 1991), and includes the period from 1985 to 1988. All values are in tons per year

Parameter (tons $year^{-1}$)	Range	Mean
Gold production	44–277	148
Hg emission factors[a]	1.3–2.0	1.5
Hg loss to the atmosphere	40–509	137
Hg loss to rivers	13–83	44
Total Hg loss	57–554	181

[a] Emission factors estimated without the use of retorts.

inputs into the Amazon region (ca. 130 tons) are comparable and even higher than the 51 tons $year^{-1}$ emitted by all industrial sources in the UK (Hutton and Symon 1986), and only 2 orders of magnitude lower than the global anthropogenic Hg emissions to the atmosphere (Pacyna 1984; Fitszgerald et al. 1984; SCOPE 1985; Nriagu 1990). The emission per area unit of Hg to the atmosphere in the Brazilian Amazon (ca. 35 g km^{-1}) is equal to the total emission per unit area resulting from industry in the USA (Fitzgerald 1994, pers. comm.).

In fact, if these numbers are used, the amount of Hg presently released in the Amazon atmosphere only can reach annually up to 5.0% and up to 2.0% of the total anthropogenic Hg emissions and total global Hg emissions to the atmosphere, respectively (Pfeiffer and Lacerda 1988; Hacon 1991). As Hg amalgamation processes in gold production have been used in the Amazon for approximately 15 years, nearly 3000 tons of Hg may have been released into the Amazon environment (Table 2.2).

Estimates of present-day emissions of Hg from amalgamation mining are listed in Table 2.3. Major impacts are from the Brazilian Amazon as described, followed by China, where in 1992 individual entrepreneurs were allowed to pan for gold, and since then over 200 small gold mining operations have been settled and 120 tons $year^{-1}$ of Hg are being released into the environment (Ming 1994; Yshuam 1994). Also of importance are the Guayana Shield region, Venezuela (40–50 tons $year^{-1}$) (Mendoza 1990; Lesenfants 1994; Nico and Taphorn 1994); the Pando Department, Bolivia (up to 30 tons $year^{-1}$ (Zapata 1994), and the Mindanao region in the Philippines with roughly 26 tons $year^{-1}$ (Cramer 1990; Torres 1992), and the Choco Department, Colombia, where from 47 semi-industrial mines, nearly 30 tons $year^{-1}$ of Hg are released into the environment. Since over

Table 2.3. Estimates of present-day annual (tons $year^{-1}$) and total emissions of Hg (tons) to the environment from gold mining sites

Site	Period starting	Annual inputs	Total input	Author
Amazon, Brazil	1979	180	3000	Pfeiffer and Lacerda (1988)
Mindanao Is., Philippines	1985	26	200	Torres (1992); Cramer (1990)
Southern Brazil	1985	1.0	10	Lacerda et al. (1995b)
Puyango River, Peru	1987	3.0	23	CIMELCO (1991)
USA	1969	6.1	152	USGS (1970)
Canada	1976	0.5	10	Hocking (1979)
Choco region, Colombia	1987	30	240	CODECHOCO (1991)
Nariño, Colombia	1987	0.5	4	Priester (1992)
Victoria Fields, Tanzania	1991	6.0	24	Ikingura (1994)
Pando Department, Bolivia	1979	7-30	300	Zapata (1994)
North Sulavesi, Indonesia	1988	10-20	120	James (1994)
Jia pi Valley, China	1938	2.4	140	Ming (1994)
Dixing region, China	1992	120	480	Yshuan (1994)
Guayana Shield, Venezuela	1988	40-50	360	Nico and Taphorn (1994)

1700 small-scale, non-industrial mines also operate in the region, the Hg released from them has not yet been evaluated; however, the total Hg input may reach very high values. In Indonesia, although involving many thousands of workers, no estimate exists for the country as a whole. However, in the North Sulavesi district, where the most intense prospecting occurs, James (1994) estimated an annual loss of metallic Hg to drainage ranging from 5 to 10 tons. Since the techniques of gold recovery in that country are similar to those in other, better-studied areas, we may suppose that a similar quantity of Hg is being released to the atmosphere. Therefore we may estimate the total input from that areas as at least 15 tons $year^{-1}$, since 1988, when consistent information is available, although this form of gold mining is traditional in that country and may date back from the last century (James 1994).

Many other small mines in North America are still operating and emitting Hg to the environment and the sum of their contributions may be significant. Also, in Africa and Southeast Asia, this environmental subject is very poorly documented, and the data from these areas in Table 2.3 are probably an under-estimation. Finally, small-scale gold mining tends to be mobile, occupying a certain area only when gold extraction through this technique is economically feasible. This mobility also makes the estimation of Hg emissions a very difficult task.

The annual input of Hg through gold mining can reach up to 550 tons, as calculated from the data in Table 2.3. This is nearly 50 % of the total Hg inputs into the biosphere mobilized by natural processes and over twice as high as the inputs from Hg mining itself. Also, these figures reach up to 5 % of the total Hg inputs from all other anthropogenic sources to the biosphere, and since industrial inputs of Hg have been effectively controlled in the last few years, and gold mining through Hg amalgamation is likely to spread throughout the world in the near future, this percentage is likely to increase.

Assuming these annual figures to be relatively constant over the last decade, gold mining may have resulted in a total release of more than 5000 tons of mercury into the environment in the last 15 years. Again, the lack of detailed documentation on gold mining using Hg amalgamation in many parts of the developing world suggeststhat this total amount may be highly underestimated (Lacerda 1995 a). Due to the relatively large residence time of Hg in superficial environments, this large amount would still cycle in ecosystems, capable of being concentrated and transported through various natural processes (Lacerda and Salomons 1991).

3 Mercury in the Atmosphere

3.1 Characteristics of Mercury Emission to the Atmosphere

Mercury emissions to the atmosphere due to precious metal mining and processing has been, until very recently, considered to be of relatively little importance compared with direct discharges into soils, tailings and rivers. However, better understanding of the whole production process, emission factors and the total amount of mercury involved has shown that Hg emission to the atmosphere from this type of mining is the most important source of this pollutant to the environment (Pfeiffer and Lacerda 1988; Hacon et al. 1990; Lacerda and Salomons 1991). It was shown in the last chapter that Hg emissions to the atmosphere represent 45 to 87% of the total Hg emitted from precious metal mining and may account for a global input of 200–420 tons $year^{-1}$. These figures represent a contribution that can reach from 6 to 11% of the global anthropogenic atmospheric emission of mercury to the biosphere of 3550 tons $year^{-1}$ (Nriagu 1990).

Atmospheric Hg may be the major limiting factor for Hg methylation in mining areas. Atmospheric deposition provides a continuous source of mercuric Hg (Hg^{2+}), necessary for the methylation process to occur (Mason et al. 1994). Therefore, even small increases in atmospheric loading could result in augmented levels of methyl-mercury in the biota (Fitzgerald et al. 1991). Metallic mercury (Hg^{0}), which is the major form of Hg mobilized in the environment through leaching of tailings and discharges into rivers from the mining operation itself, is practically non-reactive in oxic natural waters (Fitzgerald et al. 1991) and does not seem to serve as a substrate for the methylation process (Mason and Morel 1993).

Mercury emissions to the atmosphere can occur through active vaporization and passive degassing. *Active* vaporization occurs during the roasting of the amalgam and during the metal purification process. The first source occurs basically at the mining site, since miners in general do not use retorts or any closed system for burning the amalgam. Therefore this Hg ends up mostly in the rural atmosphere. On the other hand, Hg emission during purification of the bullion (where Hg appears as impurities at a concentration typically varying from 1 to 7%) (Farid et al. 1991) occurs in the shops of gold and silver dealers (located in towns) before commercialization. Here, Hg is emitted into the urban atmosphere and may be an important factor in the health of the population.

Passive degassing occurs from any contaminated soil or water body, and on a minor scale during the entire mining process due to the frequently very high temperatures of mining sites in the tropics. However, contaminated tailings are by far the most significant source of Hg through degassing to the atmosphere. Even very old tailings, deposited at least 150 years ago, are still a very important source of atmospheric Hg through degassing, resulting in anomalously high concentrations of Hg in the local atmosphere (Lechler 1993).

3.1.1 Active Mercury Vaporization: Burning of Amalgam

Burning of Hg amalgam or purification of the extracted precious metal bullion may produce various Hg species such as Hg vapor (Hg^0), HgO, Hg^{2+}, and Hg within or condensed onto particles. Particles below 1 μ diameter are formed in a gas-to-particle conversion process, while larger particles are probably formed during the roasting process itself (Haygarth and Jones 1992).

At the temperatures used in roasting of the amalgam, oxidation is also likely to occur, releasing ionic mercury followed by rapid condensation onto particles. At temperatures above 200 °C, Hg^0 is oxidized to HgO in the presence of oxygen (Hall et al. 1991). Ionic and particulate Hg have been estimated to make up 50% of the total Hg content in flue gases (Lindqvist et al. 1991). Therefore oxidation during roasting will be a significant source of reactive Hg to the atmosphere (Mason and Morel 1993). Once emitted to the atmosphere the atmospheric transport of mercury will depend on its chemical properties and on the chemical and physical characteristics of the atmosphere (Fig. 3.1).

An example of the complexity of the chemical transformations going on during during the emission process due to active vaporization was reported by Marins and Tonietto (1995). These authors investigated a selective sampling of contaminated air inside the shops of gold dealers in central Brazil mining areas. Table 3.1 summarizes their findings. Total Hg content was determined through bubbling air samples through a solution of potassium permanganate, which collects and oxidizes the total Hg present in the air sample, both vapor and particulate. Simultaneously, they

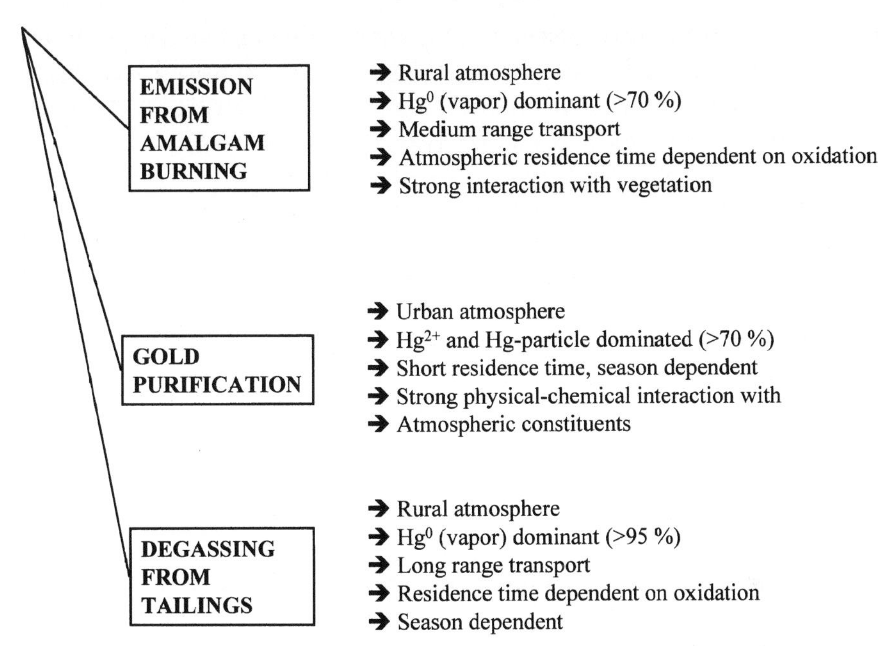

Fig. 3.1. Atmospheric mercury pathways in areas of gold and silver mining

collected air samples through deposition onto a gold filament, which provides the amount liberated as Hg vapor. The results obtained using these two different techniques showed that the concentrations, independent of the method used, were variable, depending on the working schedule of the shop, the amount of gold being commercialized at the time of the sampling, and the architecture of the shop. For example, when the shop was not operating, Hg concentrations were generally one order of magnitude lower (Table 3.1). In general, however, Hg concentrations in air samples from the shop were extremely high, frequently surpassing the maximum permissible concentration for occupational exposure of 50 μg Hg m^{-3}

Table 3.1. Mercury concentrations (μg m^{-3}) in air samples collected in a gold dealer shop using two different sampling techniques. (After Marins and Tonietto 1995)

Sampling technique	Shop condition	Hg concentration ranges
Absorption in $KMnO_4$	Before operation	6.8
(particulate and gaseous mercury)	In operation	19.7 – 106.5
Amalgamation with gold	Before operation	< 3.0 – 4.0
(gaseous mercury only)	In operation	< 3.0 – 12.0
	After closing	4.0 – 14.0

(WHO 1976). For comparison, most of the values obtained in that study are orders of magnitude higher than those found in dental workplaces (up to 3.6 μg m^{-3}) and similar to the values found in chlor-alkali plants and emissions from incinerators (Otani et al. 1986; Nilsson et al. 1990).

Mercury concentrations obtained with amalgamation with gold filament in a sniffer-type sampler showed the lowest values when sampling was done before the shop opened, ranging from <3.0 to 4.0 μg Hg m^{-3}. The maximum values were measured when the shop was operating and ranged from <3.0 to 12.0 μg Hg m^{-3}. Just after the closing of the shop Hg values were still high, ranging from 4.0 to 14.0 μg Hg m^{-3}.

Mercury concentrations using absorption in $KMnO_4$ solution also presented lower values when the shop was closed (6.8 μg Hg m^{-3}), and the highest values ranging from 19.7 to 106.5 μg Hg m^{-3}. The lowest values were measured when only small amounts of gold were commercialized.

The results using absorption in permanganate solution showed minimum concentrations when the shop was closed (6.8 μg m^{-3}) and the maximum when the shop was in full operation (106.5 μg Hg m^{-3}). Therefore, much higher concentrations than those obtained with amalgamation onto gold filament were obtained, suggesting that particulate Hg may correspond to the majority (up to 90%) of the emission.

Inside gold dealer shops the bullion is burnt before commercialization. The burning is frequently done in fume hoods equipped with exhaustion and/or retention systems for Hg to avoid local contamination. However, close to the fume hoods, air temperatures are too high. Also, during the process, smelting substances are used to purify the gold; these substances may generate oxidizing gases which may interfere with the amalgamation process in the sniffer (Driscoll 1974). Most probably, however, other Hg species are being released during the processes which are unable to amalgamate with gold. Other chemical species could possibly be inorganic mercury (Hg^{2+}) and/or non-characterized mercury species associated with particles and/or aerosols (Schroeder 1982). These would not alter the results with the two other procedures. Previous studies done in urban atmosphere of gold mining sites have shown that Hg concentrations drop very rapidly a few hundred meters from gold dealer shops (Marins et al. 1990; Malm et al. 1991). This distribution pattern of Hg concentrations in the air would be best explained if particulate Hg is a significant fraction of the total Hg emitted.

Hacon et al. (1995) collected total aerosol particulate matter (using stacked filter units) from the chimneys of gold dealer shops during the gold purification process. They found very high Hg concentrations in the particulate form. The coarse particulate mode (2–10 μm) presented higher Hg concentrations of 12.7 μg m^{-3}, whereas the fine particulate mode

(< 2 m) presented less mercury with concentrations of 2.01 μg m^{-3}. This particle size distribution may be characteristic of the gold purification process, since in the open atmosphere, other authors reported that Hg binds equally to particles of 0.4 to 10 μ in diameter (Iverfeldt 1991).

Coarse particulate represents over 86 % of the total Hg emission from those shops. Particulate Hg concentrations collected nearly 1.0 km from the shop showed only 5 to 20 % of the total atmospheric concentration of Hg, suggesting fast removal of the particulate phase very close to the source. Data on Hg deposition on urban soils close to these sources confirm a fast removal of Hg-enriched particles (Rodrigues 1994; see also Sect. 3.4).

3.1.2 Passive Mercury Vaporization: Degassing from Contaminated Soils, Waters and Tailings

Emission from contaminated soils, waters and tailings involves the degassing of Hg vapor to the atmosphere. The fluxes of Hg^0 have an important geochemical significance, since Hg can reach up to 15 % of the atmospheric deposition over natural areas (Vandal et al. 1993). Tailings (Fig. 3.2) are by far the most important site of Hg degassing, since they can reach Hg concentrations of over 5000 μg g^{-1}. In certain area such as the Folsom Canal, California, degassing from tailings is a potential source of Hg to humans living in urban developments built on tailings left from gold mining from the last century (Prokopovich 1984).

Mercury degassing rates from natural, non-contaminated areas are in general less than 10 – 20 ng m^{-2} day^{-1} or 1 – 3 μg m^{-2} $year^{-1}$ (Fitzgerald et al. 1991; Table 3.2). For example, degassing of mercury from the remote equatorial Pacific Ocean ranges from 0.43 to 6.5 μg Hg m^{-2} $year^{-1}$ at a wind

Fig. 3.2. Tailings from gold mining in central Brazil

Table 3.2. Mercury degassing rates from diverse surfaces of the Earth compared with contaminated areas from gold mining sites

Surface type	Degassing ($\mu g\ m^{-2}\ year^{-1}$)	Author
Non-contaminated soils	3.2–9.5	Rasmussen (1994)
	1.0–3.0	Fitzgerald et al. (1991)
Soils over mineralized sites	9.5–56.0	Fitzgerald et al. (1991)
Over temperate lakes	26–87	Schroeder et al. (1992)
Over the open ocean	0.3–6.5	Baeyens et al. (1991)
Over contaminated soils	130–1400	Kim et al. (1993)
Over Cinnabar ore, Almadén	2600	Ferrara and Maserti (1994)
Over last century tailings	180	Lechler (1993)

speed of 2.8 m s^{-1}, but may increase to 10.3 and 156 μg Hg m^{-2} $year^{-1}$ at a wind speed of 54 m s^{-1} (Baeyens et al. 1991). Wind speed is a key factor controlling the degassing of Hg from natural soils and water surfaces, thus the absence of plant cover on tailings may increase Hg degassing. Over mineralized areas containing cinnabar deposits, degassing rates can reach up to 60 μg Hg m^{-2} $year^{-1}$ (Rasmussen 1994).

Contaminated soils and tailings tend to show degassing rates linearly proportional to Hg concentrations in the first 15 cm of soil (Lechler 1993), similar to soils over geological anomalies (Rasmussen 1994). Surface temperature also exhibits a very strong influence on Hg^o degassing from contaminated soils. At the Carson River tailings, Nevada, USA, Lechler and Miller (1993) estimated Hg flux to the atmosphere to be over 500 ng m^{-2} day^{-1} (ca. 180 μg m^{-2} $year^{-1}$) (Table 3.2), finding a proportion between degassing rates and Hg concentration in tailings close to linearity from concentrations ranging from 0.02 to 350 μg Hg g^{-1} dry wt. of tailings. Similarly high degassing rates were measured over contaminated soils with a Hg content typical of gold mine tailings (ca. 3000 μg g^{-1}), in Oak Ridge, Tennessee (Kim et al. 1993).

Mercury speciation in the degassing flux from tailings is significantly different from emissions from natural, non-contaminated sites. Whereas natural degassing from soils typically presents a organo-Hg component, which may represent up to 10 to 20% of the total (Mora et al. 1994), over tailings this fraction is negligible, due to the low biological activity typical of tailings and the high concentration of residual metallic Hg^o. In a study of gold mining tailings in central Brazil, Tumpling et al. (1993a,b) showed total Hg concentrations in air just over tailings to range from 0.076 to 0.085 μg m^{-3}, with only

0.003 to 0.006 μg m^{-3} (less than 0.1%) of organo-mercurials. In the same region and over the same geological background, Hg concentrations in air over non-contaminated soils averaged 0.0028 μg m^{-3}, with 0.0004 μg m^{-3} (ca. 15%) of organo-mercurials (Tumpling et al. 1993a, b). However, this evidence is only preliminary and detailed studies on Hg speciation in the degassing flux from tailings should still be performed.

3.2 Mercury Concentrations in Air of Urban and Rural Areas

Table 3.3 shows mercury concentration in urban and rural atmospheres of gold mining areas worldwide. In all areas, Hg concentrations are higher than background values from non-contaminated sites and also higher than values measured in industrialized areas. Most values are also higher than those found in the atmosphere over cinnabar deposits.

The highest values were found in the urban atmosphere of towns where gold is largely commercialized, therefore where large-scale purification of gold occurs. These values, such as those reported for the cities of Poconé and Porto Velho, are orders of magnitude higher than those reported for

Table 3.3. Atmospheric Hg concentrations measured in urban and rural areas under the influence of gold mining compared with other gold producing and industrialized areas

Location	Hg (μg m^{-3})	Author
Davao del Norte, Philippines	13.5–136.4	Torres (1992)
Poconé City, central Brazil	$<$ 0.14–1.68[a]	Marins et al. (1991)
Over tailings, central Brazil	0.08	Tumpling et al. (1993a, b)
Porto Velho City, W. Amazon	0.10–3.20[a]	Pfeiffer et al. (1991)
Porto Velho City, W. Amazon	0.45–7.50[a]	Malm et al. (1991)
Humaitá City, central Amazon	0.02	Pfeiffer et al. (1991)
Madeira River mining sites	10.0–296.0[b]	Malm et al. (1991)
Over old tailings, Nevada, USA	0.01–0.04	Lechler and Miller (1993)
Over tailings, Dixing, China	14.9	Yshuam (1994)
Over Comstock tailings, USA	0.23	Gustin and Leonard (1994)
Alta Floresta City, S. Amazon	0.02–5.79[a]	Hacon et al. (1995)
Teles River mining site, S. Amazon	0.01–3.05[b]	Hacon et al. (1995)
Rio de Janeiro City, SE Brazil	0.02–0.07	Pfeiffer et al. (1989b)
Over cinnabar ores	0.01–0.09	Ferrara et al. (1982)
Background, rural values	0.001–0.015	Kothny (1974)
Background, urban areas	0.005–0.050	GESAMP (1986)

[a] Samples collected near gold dealer shops and re-burning areas in towns.
[b] Samples collected in rural areas over amalgam burning sites.

the urban atmosphere in large industrialized cities like Rio de Janeiro (Pfeiffer et al. 1989b).

In rural areas close to areas where amalgam is burned, Hg concentrations in air samples are in the same order of magnitude, ranging from 10 to 296 $\mu g\ m^{-3}$ in areas where retorts are used to minimize Hg emissions. However, at sites where no retort is used, these values are much higher, ranging from 42 to 59600 $\mu g\ m^{-3}$ (Malm et al. 1991). These high values, however, characterize typical workplace conditions.

Interesting are the concentrations found over contaminated tailings in the Carson River Valley which, although deposited in 1840s, still generate atmospheric Hg concentrations in the same order of magnitude of present-day mining sites in the Amazon.

Unfortunately, data on Hg concentrations in the atmosphere of other mining areas outside the Amazon region are extremely scarce, thus preventing a general comparison of the different mining regions presently in operation in the tropics.

3.3 Chemical Forms, Reactivity and Fate of Hg in the Atmosphere

As shown in Tables 4.1 and 4.2, Hg concentrations in the atmosphere of mining areas can be extremely high. However, as important as the total Hg concentrations are the different chemical forms of Hg in the emission, since it will eventually control the dispersal pattern, residence time in the atmosphere, and atmospheric deposition.

The four most important chemical species of Hg in the atmosphere are: elemental mercury (Hg^0), characterized by a high vapor pressure, low water solubility, and an atmospheric residence time which varies from days to a few years; divalent inorganic mercury (Hg^{2+}), with a high affinity for organic and inorganic ligands, in particular those containing sulfur radicals, particulate Hg, and methyl-mercury (CH_3Hg^+), which is highly resistant to environmental degradation, being very slowly degraded by living organisms (Lindqvist and Rhode 1984). The gaseous dimethyl-Hg $(CH_3)_2Hg$ may also be important under certain environmental conditions (Quelviviler et al. 1992). In particular over pristine environments such as the open ocean and Antarctica where, due to biological emissions and absence of antropogenic sources, dimethyl-Hg can represent up to 10% of the total atmospheric Hg concentration (Mora et al. 1993). This Hg species, however, is highly unstable under normal atmospheric conditions, rapidly degrading to methane and Hg^{2+}.

Elemental mercury is reported to account for over 95% of the total mercury in the atmosphere (Bloom and Fitzgerald 1988). This species is uniformly distributed in the troposphere with concentrations ranging from 1.0 to 2.0 ng m^{-3}. In areas under the influence of anthropogenic emissions, such as those involving high temperature processes, such as pyrometallurgy or gold purification, significant amounts of particulate-Hg may be generated, which can constitute between 5 to 20% of the total atmospheric Hg (Pacyna and Münch 1991; Hacon et al. 1995). Models of environmental mercury cycling have demonstrated that most mobilization processes involving elemental mercury and its main chemical reactions to its ionic form occur in the atmosphere.

Elemental mercury vapor released into the atmosphere during the roasting of the amalgam and vaporized during the different processes of gold extraction is oxidized to Hg^{2+} through reactions mediated mostly by ozone. Solar energy and water vapor may also participate in the oxidation process. The reaction in a pH range of 4 to 7 is linear, following the equation (Brosset and Lord 1991):

$$Hg^{o} + O_3 + 2H = Hg^{2+} + O_2 + H_2O \quad (1)$$

Therefore, rates of Hg^{2+} production will be a function of the ozone concentration in most atmospheric environments as long as a constant supply of Hg^{o} is present.

Once formed, ionic mercury (Hg^{2+}) is removed from the atmosphere by rain and is deposited in terrestrial and aquatic environments, where it may undergo further reactions including organification and assimilation by the biota (Kothny 1984; Lindqvist et al. 1984; Lindqvist and Rhode 1984; Petersen et al. 1989). Figure 3.3 summarizes the atmospheric Hg pathways in precious metal mining areas using amalgamation.

Many studies on the cycling of Hg in the atmosphere give a residence time of Hg^{o} ranging from a few months to 1 or 2 years (Slemr et al. 1981; Fitzgerald et al. 1984; Lindqvist et al. 1984). The residence time of mercury vapor in the atmosphere will depend on the rate of the oxidation process and on the adsorption of Hg^{o} onto particles. Data from many studies in gold mining areas on deposition rates (Lacerda et al. 1991b), distribution in surface soils affected by point sources of Hg (Rodrigues 1994; Malm 1992) and on the speciation of Hg in the emission from point sources (Marins and Tonietto 1995; Hacon et al. 1995) suggest much shorter residence times, of the order of days (Mason and Morel 1993).

Most tropical countries utilize slash and burn as the major agricultural practice. Forest burning increases ozone concentration in the lower atmosphere during the mining season (dry season) and has been extensively reported. For example, in the Brazilian Amazon, ozone concentrations increased from background values of 10 to 25 ppbv to 80 and 120 ppbv (Kaufman et al. 1992). Since the oxidation reaction is proportional to the

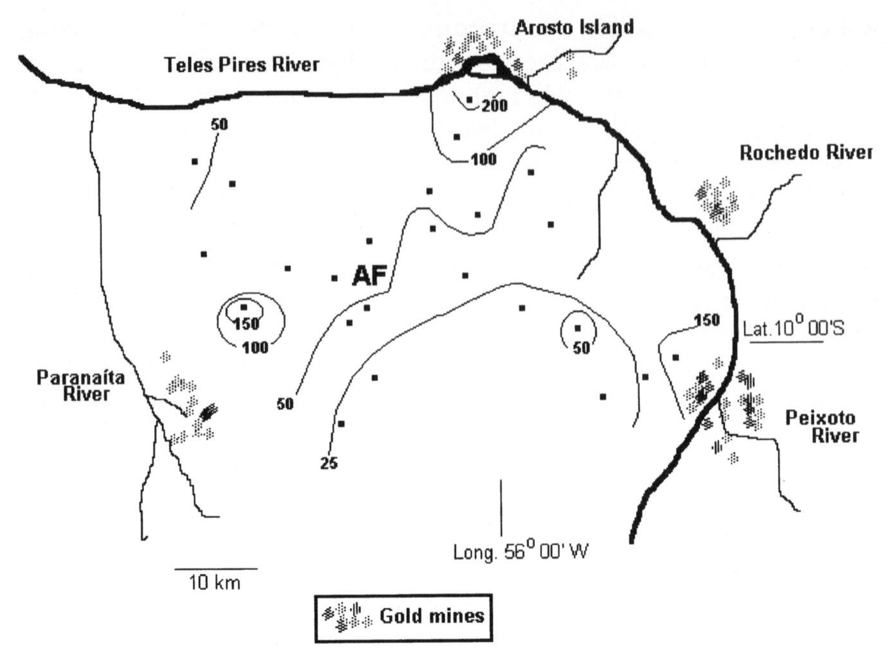

Fig. 3.3. Mercury distribution in forest soils in the Alta Floresta, Mato Grosso State

ozone concentration [Eq. (1); Brosset and Lord 1991], the formation of Hg^{2+} in the atmosphere would be very fast during the mining (dry) season.

Another important control of the residence time of Hg in the atmosphere is the concentration of suspended particles and aerosols. During forest burning, particle loading increases from 10 to 20 μg m^{-3} to values as high as 700 μg m^{-3} (Artaxo et al. 1993). Analysis of inorganic constituents of Hg-enriched atmospheric particles showed that soil dust and coal resulting from the burning of vegetation are efficient traps of mercury vapor present in the atmosphere and may also serve as oxidation sites in the production of Hg^{2+}. Soil dust particles collected in gold mining towns showed consistently high Hg concentrations of up to 1554 μg g^{-1} of Hg and an average of 31 μg g^{-1} of dust (Thornton et al. 1994).

The increase in ozone concentration and in particle load will therefore result in short Hg residence times in the atmosphere, due to rapid oxidation of gaseous Hg, which would then be easily washed out by the frequent rains occuring in most tropical regions.

3.4 Impacted Areas Through the Atmospheric Transport Pathway

3.4.1 Soils

Soils present special integration characteristics due to their low mobility and transport. They are good indicators of atmospheric deposition, being a permanent or temporary sink but also a source of volatile Hg species. Contaminated soils can be a significant source of Hg to vegetation. Mobility of Hg in the surface soil is controlled by the strong affinity of the bivalent form (Hg^{2+}) for organic matter; the degassing of the elemental form (Hg^{0}) and by the adsorption–desorption reactions of these two forms (Rasmussen 1994). In general, Hg mobility in soils is extremely low, making this compartment an important tool for monitoring Hg.

Considering the potential large-scale dispersion of atmospheric Hg in gold mining areas, it is expected that soils around mining sites should also be contaminated. However, notwithstanding the importance of this compartment for the understanding of Hg cycling, few studies deal with this compartment.

3.4.1.1 *Urban Soils*

Data on Hg concentrations in urban soils from gold mining regions (Table 3.4) have mostly been obtained from studies on gold purification through high temperature roasting in gold dealer shops (see Chap. 4). In general, all studies on urban soil contamination in the Amazon region showed that Hg concentrations decreased rapidly from its source (Marins and Tonietto 1991), and this is probably due to particle deposition in the vicinity of the source (Marins et al. 1990, 1991; Hacon et al. 1994). Soils collected in urban areas of the same region where roasting of gold with Hg impurities occurs presented extremely high Hg concentrations, ranging

Table 3.4. Mercury concentrations in urban soils affected by gold dealer shops and workplaces

Site	Hg ($\mu g\ g^{-1}$)	Author
Porto Velho, Rondônia, close to gold dealer shops	0.46 – 64.0	Malm (1993)
Porto Velho, Rondônia, far from gold dealer shops	0.03 – 1.33	Malm (1993)
Alta Floresta, south Amazon	0.05 – 4.1	Farid (1992)

from 0.46 to 64 μg g^{-1} close to gold dealers shops, to 0.03 to 1.3 μg g^{-1} dry wt. only 300 m from those shops (Malm et al. 1991).

Extremely high Hg concentrations have been found in urban street dust. In the town of Porto Velho, Madeira River area, Malm et al. (1991) found Hg concentrations up to 36 μg g^{-1} of dry dust, while Thornton et al. (1991) found up to 250 μg g^{-1} of dry dust in the mining village of Itaituba, southern Amazon. In a study of the characterization of Hg species emitted by roasting of gold in the shops of gold dealers, Marins and Tonietto (1991) suggest that particulate Hg is the major species of Hg emitted rather than Hg vapor, thus explaining the distribution of Hg found in urban soils.

Analysis of urban soil samples, collected all on the same day in the vicinity of reburning areas in the city of Porto Velho, SW Amazon, indicates a dispersion circle of around 600 m as well as the effect of prevailing winds on Hg deposition (Malm et al. 1991). Data showed that the proximity of gold shops strongly determines the Hg soil concentrations, independently from origin (different cities, forest or urban soils).

Urban soils in Alta Floresta, Brazil, showed the importance of Au purification in contributing to high levels of Hg in soils. Mercury concentrations ranged from 0.05 to 4.10 μg g^{-1}, with a mean of 0.23 μg g^{-1} (20% of the samples were $<$ 0.10 μg g^{-1}; 54% from 0.10 to 0.20 μg g^{-1}; 15% from 0.2–0.3 μg g^{-1}; and 11% of the samples with Hg concentrations $>$ 0.3 μg g^{-1} (Rodrigues 1994). The highest concentrations were found up to 1000 m from the nearest source.

No relation was found between Hg concentrations and organic matter content, at least for urban soils, neither in surface soil nor in core samples, reinforcing the idea that source proximity and particle transport and deposition patterns are more important than chemical ones for explaining dispersion and retention of Hg, especially when it is being emitted in the metallic form in urban regions.

3.4.1.2
Forest Soils

Forest soils have received more attention than urban soils, although they present in general much lower Hg concentrations. A summary of results on Hg concentrations in tropical forest soils is presented in Table 3.5.

Forest soils from the Madeira River watershed, northwestern Amazon, were analyzed by Lacerda et al. (1987) and showed Hg concentrations ranging from 35 to 300 μg kg^{-1} dry wt. The highest values were found close to river sections where intense mining takes place. Malm et al. (1991) also studied forest soils in the Madeira River area. These authors found similar results, with Hg concentrations ranging from 0.03 to 0.34 μg g^{-1} dry wt. in

Table 3.5. Distribution of mercury in forest soils in areas with mining sites

Site	Hg (µg g^{-1})	Author
Forest soil at Tucuruí, pristine forests, SE Amazon	0.094–0.13	Aula et al. (1994)
Forest soil in French Guyana, inundated pristine equatorial forest	0.15–0.28	Roulet and Lucotte (1994)
Forest soil in Pará, pristine forest SE Amazon	0.120–0.20	Lucotte (1994)
Forest soils, pristine, Rondônia, W. Amazon	0.03–0.34	Malm (1993)
Forest soil, pristine, Teotônio area, Rondônia, W. Amazon	0.035–0.30	Lacerda et al. (1987)
Forest soils close to mining sites, Madeira River basin, Rondônia, W. Amazon	0.42–9.99	Malm (1993)
Forest soils of active working areas, Bolivar, Venezuela	1.43–129.3	Sherestha and Quilarque (1989)
Forest soils of active mining areas in Victoria Gold fields, Tanzania	0.050–28.2	Ikingura (1994)
Forest soils from Serra Pelada, SE Amazon	0.29	Araújo Neto (1990)

forest soils relatively far from mining sites. In forest soils close to amalgam burning areas, Hg concentrations were much higher, ranging from 0.42 to 9.9 µg g^{-1} dry wt. They also found a 100% increase in Hg concentrations in surface soil when compared to deeper horizons in soil profiles. They associated high mercury content in the topsoil with atmospheric deposition and with the high (up to 27%) organic matter content of these soils.

Strikingly large differences in mercury concentrations in adjacent soils in use as pastures or covered by forests were found in a recent study in the Alta Floresta, Mato Grosso State. The soils from this area cover an area of ca. 5000 km^2. Figures 3.3 and 3.4 show the location of mining areas and the contours of the mecury distribution in the soils.

Mercury concentrations in surface soils varied according to the distance from Hg sources in forest and pasture soils. Mercury concentrations ranged from 37 to 210 µg kg^{-1} in forest soils and from 17 to 42 µg kg^{-1} in pasture soils. The highest Hg concentration in forest soils ranged from 150 to 210 µg kg^{-1} and from 30 to 42 µg kg^{-1} in pasture soils. Mercury concentrations in forest soils are in the range reported for other areas in the Brazilian Amazon and in French Guyana (Aula et al. 1994; Roulet and Lucotte 1995), but were lower than the concentrations reported for urban soils of Alta Floresta (Rodrigues 1994).

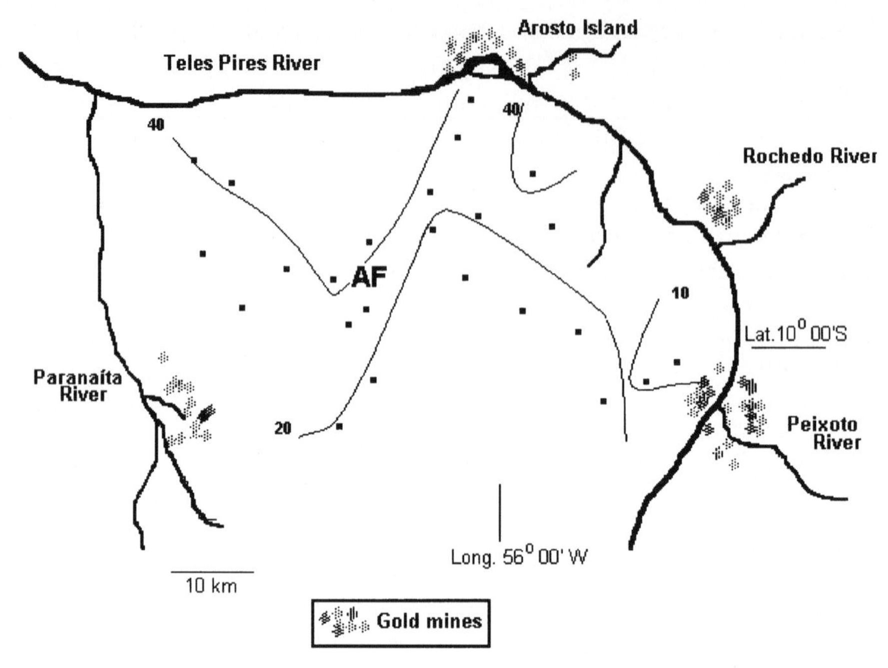

Fig. 3.4. Mercury distribution in pasture soils in the Alta Floresta, Mato Grosso State

The influence of the proximity of Hg sources for forest and pasture soils is clearly shown in the figures. The distribution of iso-concentration lines suggests that Hg emitted from mining sites to the atmosphere is deposited in the first 10 to 20 km from the source. Mercury deposition relatively close to sources has been also reported for urban areas, however, with much higher concentrations and within a maximum 1.0-km radius from the source (Marins et al. 1991; Rodrigues 1994).

Mercury concentrations were significantly higher in forest soils than in pasture soils ($P < 0.01$, n = 31). Notwithstanding the clear influence of source proximity in forest soils, this influence is much less pronounced in the distribution of Hg in pasture soils, suggesting that Hg loading rates in pastures soils is not reflected in Hg concentrations. Since the two soils are similar in composition (yellow-red latosols), any difference in background concentrations of mercury can be ruled out as an explanation. The differences are caused by differences in vegetation, exerting two effects. The forest with its higher specific surface area has a higher efficiency for trapping the atmospheric deposition. The second effect is probably due to the difference in the soil temperature. which is expected to be higher in pasture areas, this will result in a high degassing rate of mercury from the soil.

Soils collected from an inundated equatorial forest in French Guyana, showed Hg concentrations ranging from 0.15 to 0.28 μg g^{-1}. These relatively high Hg concentrations were attributed to atmospheric deposition and accumulation during the humification process (Roulet and Lucotte 1994).

In another study in the Poconé region, central Brazil, soils samples from an approximately 10 000 km^2 area under the influence of various mining sites showed very low Hg concentrations, being smaller than 0.03 μg g^{-1} dry wt. (which is considered the local background Hg concentration), in 70% of the analyzed samples. In circa 30% of the samples, Hg concentrations ranged from 0.03 and 0.1 μg g^{-1} dry wt. Close to mining sites however, Hg concentrations reached 0.27 μg g^{-1} dry wt. (Lacerda et al. 1991a). Compared to European and Canadian soils, with Hg concentrations between 0.07 and 0.3 μg g^{-1} dry wt. (Mitra 1986), the values found in this region are relatively lower. However, if one considers the local background Hg concentrations in Amazon soils, Hg contamination is very clear, at least close to mining sites.

A consistent trend of increasing Hg concentrations with depth has been found in some tropical forest soils. Lucotte (1994), working along the Tapajós River Basin, SE Amazon, found an increase in Hg concentrations from 0.10 μg g^{-1} at the surface litter to 0.12 μg g^{-1} in the humus layer, to 0.20 μg g^{-1} at 30 cm depth. Aula et al. (1994), working at the Tucuruí Reservoir basin, found 0.094 μg g^{-1} at the soil surface and 0.13 μg g^{-1} below 50 cm depth (Fig. 3.5). Roulet and Lucotte (1994) found a two- to threefold increase in Hg concentrations from the litter layer (0.08 – 0.15 μg g^{-1}) to the humus layer (0.25 μg g^{-1}) in soils from an inundated forest in French Guyana. These results suggest an enrichment of Hg during the humifica-

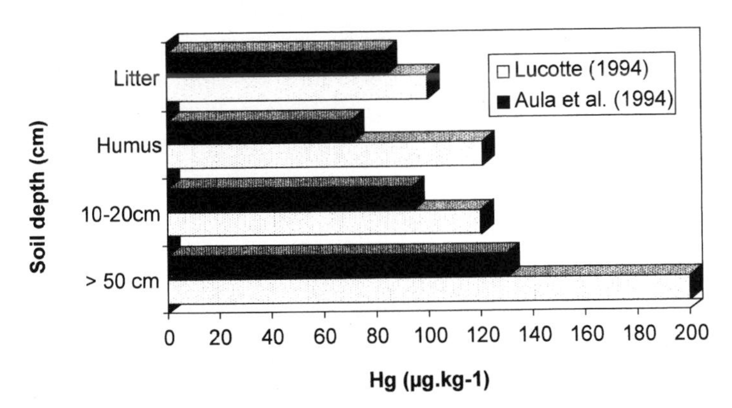

Fig. 3.5. Mercury distribution in different soil depths from forest areas in the southeastern Amazon (Aula et al. 1994) and from French Guyana. (Lucotte 1994)

tion process. Since Hg is exceptionally tightly bound by humates, humification processes would relatively increase the Hg content by losing other organic and inorganic constituents. Other accumulative mechanisms of Hg in the humus layer include biotic and abiotic conversion of inorganic Hg to less mobile organic complexes and volatilization from deeper layers (Rasmussen 1994). However, a significant increase from surface to deeper mineral soil was also observed (Aula et al. 1994; Lucotte 1994; Fig. 3.5). Since the geological background of these tropical areas shows very low Hg concentrations (0.01 to 0.03 μg g^{-1}; Lacerda et al. 1991c), the only explanation for the observed trend is a slow, long-term migration of Hg from surface horizons in these very old and weathered soils (Lucotte 1994).

3.4.2 Lakes

Mercury which enters a lake accumulates ultimately in the bottom sediments. Analysis of the development of concentrations of metals along sediment cores, therefore, makes it possible to determine the history of metal contamination for a certain region, as well as providing important information regarding background Hg concentrations (Aston et al. 1973; Hakanson 1974; Förstner and Wittman 1981; Förstner and Salomons 1983; Scrudato et al. 1987).

Atmospheric Hg deposition rates over remote areas not affected by direct anthropogenic sources are in general less than 20 μg m^{-2} $year^{-1}$ (Fitzgerald et al. 1991; Glass et al. 1991; Lindqvist et al. 1991), and can reach values 10 to 100 times lower in extreme remote sites such as Antarctica (Vandal et al. 1993). For contaminated areas, typical deposition rates may range from 40 to 50 μg m^{-2} $year^{-1}$ in areas where contamination comes from regional diffusive sources, such as South Florida (Mason and Morel 1993) to 160 to 370 μg m^{-2} $year^{-1}$, for heavy industrialized areas, such as in northern Europe and the USA (Madsen 1981; Rekolainen et al. 1986).

Mercury atmospheric deposition rates from gold mining areas has received very little attention and was only studied indirectly, through Hg concentrations in lake sediment cores in central Brazil (Lacerda et al. 1991).

Figure 3.6 shows Hg distribution in sediment profiles of two lakes in which drainage does not receive direct contaminated effluents from mining, It can therefore be assumed that variation in Hg concentrations through the cores is due to variation in the Hg atmospheric deposition rates. As gold mining activities in the region, and therefore Hg emissions to the atmosphere, started in 1981, the distribution of Hg throughout sediment profiles can be used to estimate sedimentation rates and Hg atmospheric deposition for the region. Since Hg background concentrations in

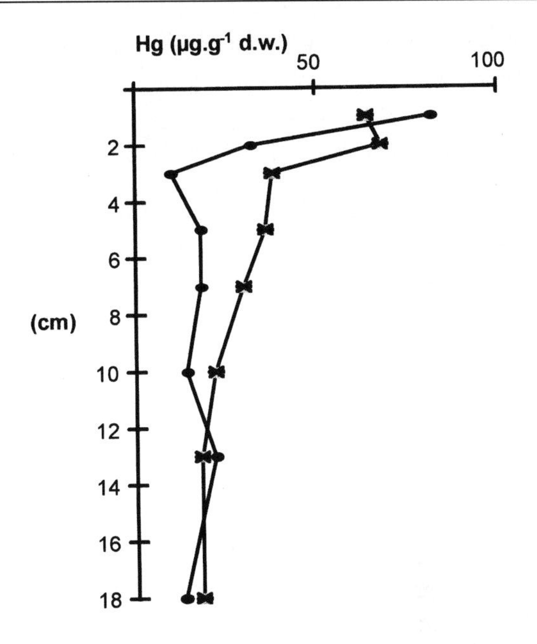

Fig. 3.6. Distribution of mercury concentrations in sediment profiles from lakes in the gold mining region of central Brazil. (After Lacerda et al. 1991b)

the area are between 10 to 30 μg kg^{-1}, which correspond to Hg concentrations from 2.0 to 30 cm of depth in the sediment core, and the concentrations at the surface of sediments reached 60 to 80 μg kg^{-1}, a steady increase in Hg concentration from around 2.0 cm of depth is clear (Fig. 3.6).

Using the Hg concentrations found in the top 2 cm (60 to 80 μg kg^{-1}) we can estimate a Hg accumulation rate ranging from 90 to 120 μg Hg m^{-2} $year^{-1}$. This estimate is quite in accordance with Hg emissions to the local atmosphere during the last 10 years. Lindqvist et al. (1991) suggested that half of the Hg in combustion flue gases is ionic in nature, both the gaseous and particulate form. The data obtained by Marins and Tonietto (1995) in the atmosphere of gold mining areas north of these lakes also suggest that elemental Hg released during the mining activity is rapidly oxidized during the roasting process and after reaching the atmosphere. This would explain the high deposition rates found and probably represent the higher range of deposition rates, since it was in the Poconé urban area that one of the highest Hg concentrations in the atmosphere was measured (see Tables 4.1 and 4.2).

Table 3.6 compares Hg deposition rates estimated for the Poconé lakes and other deposition rates published in the literature and estimated from sediment core data. Although Hg concentrations in surface and deep sediments of the Poconé lakes discussed above are lower than those values reported for industrialized areas, deposition rates are very similar to such areas, suggesting the importance of atmospheric Hg contribution from

Table 3.6. Mercury in sediment profiles from various lakes and estimated deposition rates

Location	Background concentration ($\mu g\ g^{-1}$)	Surface sediments ($\mu g\ kg^{-1}$)	Deposition rates ($\mu g\ m^{-2}\ year^{-1}$)	Author
Poconé, Brazil	15–25	60–80	90–120	Lacerda et al. (1991b)
Finland, remote	20–50	20–50	25–50	Rekolainen et al. (1986)
Finland	50–250	170–550	370	Rekolainen et al. (1986)
Denmark	151–523	243–314	30–200	Madsen (1981)
Finland	250	430	120	Simola and Lodenius (1983)
Norway	15–40	60–190	9–30	Steines and Anderson (1991)

gold mining. These high deposition rates were confirmed by direct measurements of bulk deposition over this same area, which also gave values as high as 87 $\mu g\ m^{-2}\ year^{-1}$ (Tumpling, pers. comm.).

Mason et al. (1994) estimated Hg deposition flux between latitude 10 °N and 10 °S as 12.7 $\mu g\ m^{-2}\ year^{-1}$. This is nearly ten times lower than the estimate for central Brazil discussed above. However, the existing results on mercury concentration and distribution in the Amazon atmosphere affected by gold mining are, however, too preliminary, hampering the development of a reasonable model for its behavior in such an important environmental compartment. The high deposition rates suggested by the few studies indicate that atmospheric Hg may be the dominant source of Hg^{2+} for methylation reactions.

4 Mercury in Tropical Aquatic and Terrestrial Systems

The fate of Hg in water, sediments and soils in gold mining areas has been studied mostly in large surveys at a great number of mining sites, few however, have been studied in detail. Among the many sites, the Amazon region, including Brazil and Venezuela, Mindanao Island in the Philippines and a few last century USA gold rush sites, has been the subject of long-term studies concerning the fate of Hg in the environment. Unfortunately, however, even in these areas, few studies have dealt with geochemical and biogeochemical processes controlling Hg concentrations in water, sediments and soils.

4.1 Characteristics of Mercury Emissions to the Aquatic Environment

The different mining processes using Hg amalgamation result in different wastes, Hg dispersal mechanisms, degree of Hg mobility, and biological availability in terrestrial and aquatic ecosystems. Figure 4.1 describes the main pathways of Hg released into soils and rivers.

In areas where gold is mined from active bottom sediments from rivers, mercury is lost to the environment as metallic Hg (Hg^o) directly into rivers (see also Sect. 1.3). Where the mining operation involves grinding of Au-rich soils, metallic Hg is concentrated in tailings and can eventually be mobilized through leaching and particle transport during rains (see also Sect. 1.3). Metallic Hg, however, is insoluble and practically non-reactive under the normal oxic conditions present in most surface environments, at least for many decades (Fitzgerald et al. 1991). Moreover, it also displays very low availability to biological uptake (Taysayev 1991).

In river mining operations, fluvial transport can move Hg considerable distances in a few years. Dispersal of Hg contained in tailings, on the other hand, can be a very slow process, sometimes involving the transport of the tailings themselves through erosion (Miller et al. 1993), and migration through groundwater (Prokopovich 1984). These releasing processes may

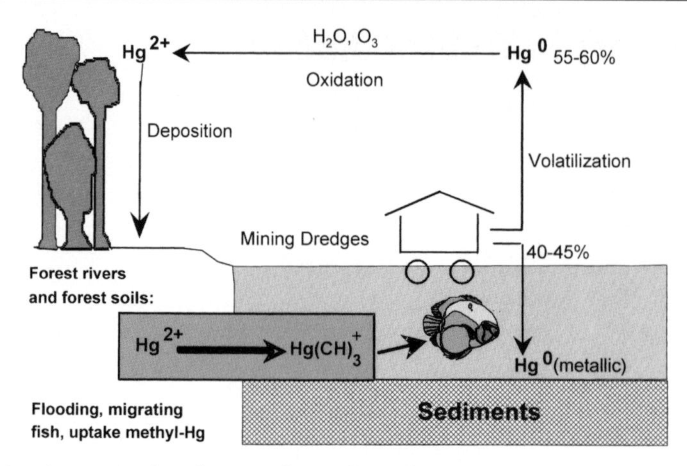

Fig. 4.1. Environmental pathways of Hg released to aquatic and terrestrial systems

last through time, posing an environmental contamination risk even after mining activity has ceased for centuries (Fuge et al. 1992). In both mining situations, a major proportion of Hg is lost to the atmosphere either through burning of the Au-Hg amalgam or through volatilization of metallic Hg from soils, sediments and rivers.

After Hg is introduced and dispersed into a river system, its behavior, no longer influenced by the source, will be affected by the general limnological conditions of tropical rivers. Mercury will be incorporated in the sediments and will also remain partly in solution. In the sediments it will not be evenly distributed over the various grain-size fractions or over the various constiutents like organic matter and clay minerals which make up the sediment. Furthermore, in the aquatic environment, mercury will be transformed (partly) into the highly toxic methyl-mercury form. The fate of mercury once introduced into the aquatic environment will depend on the charactristics of the receiving waters.

4.2 Leaching from Tailings

4.2.1 Mercury Distribution in Gold Mining Tailings

In areas where gold is mined from soils or rock, Hg is deposited and accumulated in mine tailings. An important question is how Hg is eventually transported to, and spread through, adjacent pristine environments. Although only few studies have dealt with this subject, a general pattern of

Hg distribution in tailings can be observed based on data gathered from old and recently deposited tailings.

- In the majority of tailings Hg is present as metallic Hg. Water-soluble Hg is generally less than 0.1% of the total Hg content in tailings (Andrade et al. 1988; CETEM 1989; Lacerda 1990; Ramos and Costa 1990; Lechler and Miller 1993). A larger fraction of Hg forms is only extractable with aqua regia, indicating that these forms are associated with residual minerals or elemental Hg^0 (Ching and Hongxiao 1985). A fraction of Hg lost to tailings may also be amalgamated with gold grains (Lane et al. 1988; Fuge et al. 1992).
- The distribution of mercury in tailings is a function of the distribution of mining activities in the area rather than any geochemical process. In areas where old pre-concentration ponds existed, Hg concentration can reach a few micrograms per gram of tailings material, while mean Hg contents in the whole tailings body are similar to local background values (Prokopovich 1984; CETEM 1989).
- Different mining techniques may result in different Hg concentrations and distributions in tailings. Typically, these concentrations range from less than 0.5 $\mu g\ g^{-1}$ up to 4900 $\mu g\ g^{-1}$ (dry weight) (Andrade et al. 1988; Ramos and Costa 1990; Lechler and Miller 1993).
- Mercury present in gold mining tailings can represent an important source of contamination even centuries after the cessation of mining operations (Lane et al. 1988; Bycroft 1992; Fuge 1992; Lechler and Miller 1993)

The four characteristics listed result in a very low reactivity and bioavailability of Hg in tailings, but turn these deposits into a long-lasting source of contamination.

Figure 4.3 shows a typical distribution of Hg concentrations in gold mining tailings in central Brazil obtained from the analysis of a 200-sample point grid (CETEM 1989) from the Tanque dos Padres tailings (Fig. 4.2). It is clear that the Hg distribution is not normal, presenting background values in over 95% of the samples ($< 0.20\ \mu g\ g^{-1}$). The few samples with anomalously high Hg content (4000 to 25000 $\mu g\ g^{-1}$) showed no distribution pattern and were not correlated with any geochemical or geological aspect of the tailings (CETEM 1989).

Prokopovich (1984) measured Hg concentrations ranging from 36–200 $\mu g\ kg^{-1}$ in over 60 samples of dredge tailings from the last century in the Folsom-Natomas gold field, California. In three samples randomly taken, however, Hg concentrations reached 37500; 21500 and 22000 $\mu g\ kg^{-1}$. If mapped, these results would give a picture similar to that in Fig. 4.3.

Fig. 4.2. The Tanque dos Padres tailings, Poconé mining region, central Brazil

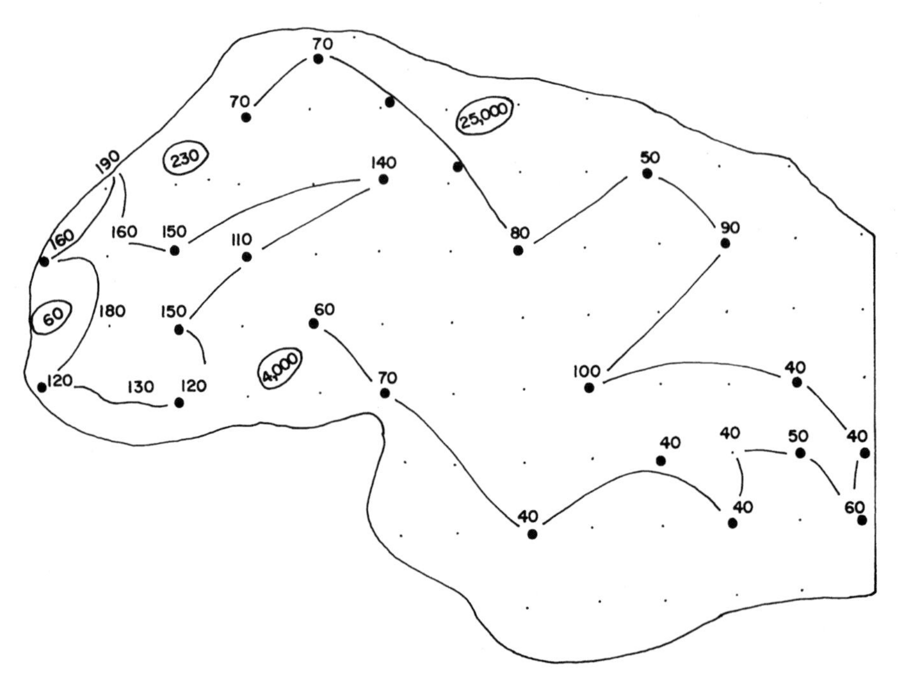

Fig. 4.3. Mercury distribution in tailings samples from gold mining operations in central Brazil (Tanque dos Padres). (After CETEM 1989)

4.2.2 Mercury Export from Tailings to Adjacent Aquatic Systems

In small creeks draining mine tailings in central Brazil, Andrade et al. (1988) found that Hg contamination of the Crixás deposit, Goiás State, decreases from a maximum value of circa 16 μg g^{-1} at the tailings to background levels of circa 0.7 μg g^{-1} in the first 200 m along the drainage. The same behavior was found by Lacerda et al. (1991a) at the Tanque dos Padres tailings, Mato Grosso State. Mercury concentrations decrease from values of up to 3.0 μg g^{-1} in the tailing to background levels (~ 0.04 μg g^{-1}) in the first 100 m along the drainage. Although Hg leaching from tailings seems a slow process, over the years, the migration of tailings themselves may be the most important Hg dispersal mechanism. In the Carlson City mining district over the past 134 years, Hg-rich tailings have been transported over 100 km from the Six Mile Canyon and tributaries to the Carlson River Valley forming a Hg-rich 1.5 km^2 fan area (Lechler and Miller 1993; Miller et al. 1993). At the Falsom-Natomas region, California, it has been suggested that groundwater migration could export Hg from 1868 dredge tailings, based on anomalous concentrations found in groundwater samples from that area. In Nova Scotia, Canada, however, groundwater samples collected in last century tailings with Hg content ranging from 1.1 to 19.9 μg g^{-1}, no contamination of groundwater could be detected (Lane et al. 1988). Creek waters, with up to 40 μg g^{-1} of Hg in the Waverly area, showed Hg concentrations always below 50 μg l^{-1} (Mudroch and Clair 1986). Surface and groundwater samples collected from 21 stations in creeks and tailings in North Carolina also showed concentrations below 20 μg l^{-1} (Callaham et al. 1994).

Strong gradients have been frequently found in both sediments and water Hg concentrations from point sources of Hg, such as tailings. In general, these gradients are very steep, indicating the low dispersability of Hg from the releasing points. Figure 4.4 shows Hg concentrations in sediments of rivers draining tailings deposits in Poconé, central Brazil (Lacerda 1991). It is clear that Hg is very refractory to transport since its concentrations drop to nearly local background values only a few kilometers from tailings.

Similar results have been found in other tailings-affected drainages in the Crixás mining site, Goiás State, central Brazil and in the Cumaru mining site, northern Amazon (Andrade et al. 1988; Ramos and Costa 1991).

Lacerda et al. (1991a) studied Hg dynamics in a creek draining a contaminated tailing deposit during a storm in the Poconé mining site, central Brazil. Significant changes in Hg concentrations in water and suspended

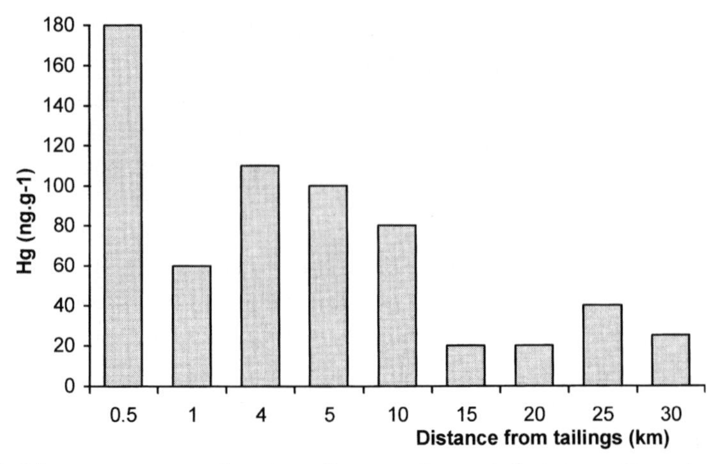

Fig. 4.4. Mercury concentrations in sediments of a creek draining tailings deposits in Pocone, central Brazil

particles, as well as in major physico-chemical parameters in creek water draining the tailings, occurred during storm events. The water became progressively more acidic with pH dropping from 6.8 to 5.1. Redox potential also increased owing to the entry of rainwater supersaturated in oxygen and average Eh became more oxidizing, increasing from 280 to 430 mV 10 min afterthe start of the storm. The content of suspended particles also showed a dramatic increase, reaching values up to ten times higher than during dry conditions, increasing from 6.0 to 66 mg l^{-1}. These changes occurred just a few minutes after the beginning of rainfall. Mercury concentrations in suspended particles increased from < 0.02 to 0.61 $\mu g\ g^{-1}$, although dissolved Hg concentrations remained below 0.04 ng l^{-1}. These results show the capacity of rain to erode fine particles enriched with Hg, from the tailings, suggesting that erosion, followed by Hg transport associated with suspended particles, is the main pathway of Hg contamination of the tailings drainage. The fact that dissolved Hg concentrations remained constant during the storm suggests that only small, if any, amounts of Hg are remobilized into solution and that Hg concentration in water is independent of mercury content in sediments (Andrade et al. 1988). Resuspension of bottom sediments from winter storms also resulted in higher particulate Hg concentrations in Clear Lake, California, a contaminated area with mine tailings (Gill and Bruland 1990).

Nriagu et al. (1992), studying Hg partitioning between dissolved and particulate forms in the Madeira River, also showed that dissolved mercury concentrations are independent of its concentrations in suspended matter, whereas Lechler and Miller (1993) showed that water-reactive Hg in tailings was so low that its concentrations could not be detected through conventional routine analytical techniques.

Mudroch and Clair (1986) found the same Hg distribution pattern in waters draining tailings in Canada. They also suggested that contamination of sediments was the result of transport and deposition of suspended particles brought into the drainage by erosion of mining wastes. Lechler and Miller (1993) also found similar behavior of Hg in waters draining tailings in the Carson River Valley, USA. Prokopovich (1984) found a significant relationship between Hg content and the clay faction in old California tailings.

The fact that none of these studies found significant changes in dissolved mercury concentrations, even during strong storm events, suggests that only small amounts of Hg are remobilized into solution and that Hg concentration in water is independent of Hg content in sediments (Andrade et al. 1988). Thus, controlling waste erosion and suspended particle transport by means of dams and impermeabilization and other protecting measures to tailings should be quite efficient in preventing Hg contamination in areas surrounding tailings.

4.3 Rivers

4.3.1 River Systems in Tropical Areas

Attempts have been made to study the influence of the different tropical river types upon the distribution of Hg, since such river types (cf. white, black and clear water rivers) present different hydrochemistries which strongly influence various other constituents of river systems in the tropics, including sediment geochemistry, trace metal distribution and aquatic biota (Furch et al. 1982; Lacerda et al. 1990).

Limnological classification of tropical rivers was first applied to the Amazon Basin by Sioli (1950). This classification finds some parallels with most other tropical regions (Payne 1986). In summary, tropical rivers have been classified into three classes based on their major hydrochemical properties by various authors (Sioli 1950; Junk and Furch 1980; Furch et al. 1982).

- "*White Water*" rivers, which are rich in suspended matter (> 200 mg l^{-1}) have a neutral pH and moderate electric conductivity (> 40 µS cm^{-1}), and mean dissolved element concentrations similar to the mean of world rivers; the Amazon River itself and the Madeira River are typical representatives of this class of river. These rivers originate in the Andes mountains where steep topography exposes sedimentary rocks, which are the major source of the high suspended solids load typical of them (Stallard 1985; Stallard and Edmond 1983). Other tropical rivers which drain geological young mountains present the same characteristics. For example, the Purari River in New Guinea and the Indus would fit into such a category (Gibbs 1972; Payne 1986).

- *"Black water"* rivers, which drain weathered, sandy tropical soils and floodplains, are rich in dissolved organic substances, are acidic (pH < 5.0) and have low concentrations of dissolved constituents (electric conductivity $< 10\ \mu S\ cm^{-1}$). The Negro River in the Amazon Basin is a typical representative of this class of river, but many rivers of the Zaire basin also fit into this category (Welcomme 1979).
- *"Clear water"* rivers present acidic to neutral pH, low organic and inorganic dissolved constituents and are relatively rich in Fe oxides from weathered laterite soils where they originate (Sioli 1950; Furch et al. 1982).

The distribution of Hg in the sediments of ten rivers of the Madeira River basin belonging to the three different classes was studied by DePaula (1989) and Lacerda et al. (1990b). In this basin, Hg is released in a sector of the Madeira River only, reaching the tributaries only through the atmosphere and during flood periods, when the Madeira River water overflows, entering the tributary river system. Figure 4.5 shows the location of the studied rivers in relation to major mining sites, while Table 4.1 summarizes the major results of those studies. The results shows that black water rivers are enriched with Hg when compared to other river classes, notwithstanding the fact that direct Hg inputs in that region are restricted to white water rivers. The enrichment of Hg in black water rivers may be related to the high organic matter content typical of this river class and to their acidity. Mercury would form relatively refractory compounds with organic matter, facilitating its accumulation in river sediments. Also, acidity would accelerate the oxidation of Hg^0 to Hg^{2+}, enhancing the possibility of Hg binding to organic matter (Lindqvist et al. 1984). Similar results have been found for other trace metals in the same rivers and were also associated with the higher organic matter content of black water river sediments (DePaula et al. 1990; Lacerda et al. 1990).

The results discussed above, however, are still preliminary, and other large tropical river basins should be tested before a generalization can be made, although the evidence available strongly suggests that the role played by organic matter is very important in the accumulation of Hg in bottom sediments of tropical rivers.

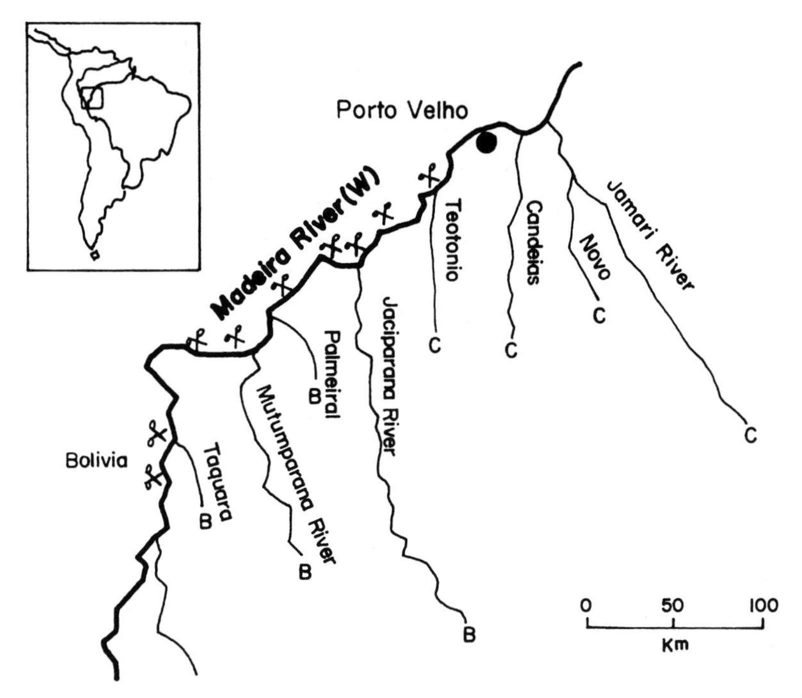

Fig. 4.5. Major tributaries of the Madeira River basin, Rondonia State, classified according to their hydrochemical class. *C* Clear water; *B* black water and location of major mining sites

Table 4.1. Major physicochemical characteristics of Amazonian rivers of different classes in the Madeira River basin, Rondonia State. (After DePaula 1989; Lacerda et al. 1990)

River class	White water	Black water	Clear water
pH	6.7 ± 0.3	5.7 ± 0.7	5.7 ± 0.6
Conductivity (μS cm^{-1})	49.0 ± 14.0	6.0 ± 1.0	13.0 ± 7.0
Sediment organic matter (%)	3.9 ± 1.2	9.0 ± 4.4	5.2 ± 2.8
Sediment Fe (%)	4.9 ± 0.8	0.5 ± 0.7	6.6 ± 2.2
Sediment Hg (μg g^{-1})	0.33 ± 0.81	0.49 ± 0.69	0.13 ± 0.08

4.3.2 Mercury and Its Associations in River Sediments

Lacerda (1990) hypothesized that Hg could also be transported associated with particulate organic carbon (POC) derived from the decomposition of plant litter brought into tropical rivers during the flooding period. Mercury dispersal associated with POC has been reported as an efficient long-range dispersal mechanism in other areas (Lindberg and Harris 1974). In the JiYum River, Ching and Hongxiao (1985) suggested that humic acids were responsible for increasing solubilization of Hg originating from upstream tailings. Melanetti et al. (1995) demonstrated experimentally the increasing solubility and decreasing adsorption onto sediments of metallic Hg in the presence of humic acids in waters draining tailings in central Brazil.

Organic matter content of sediments has also been claimed to control Hg accumulation in freshwater sediments (Rekolainen et al. 1986). In Brazilian gold mining areas, for example, the relationship between organic matter and Hg accumulation was studied in five remote lakes in the Patanal area, central Brazil, far from direct inputs of mercury from gold mining. In these lakes Hg inputs originate solely from the atmosphere (Lacerda et al. 1991c). Figure 4.6 shows the relationships between Hg and total iron and organic matter content in surface sediments of these remote lakes. No significant relationship has been found with iron. However, a significant positive relationship was found between Hg and organic matter ($r = 0.82$; $P < 0.01$). Therefore, these results seem to corroborate those found in black water rivers in the Madeira River basin described above.

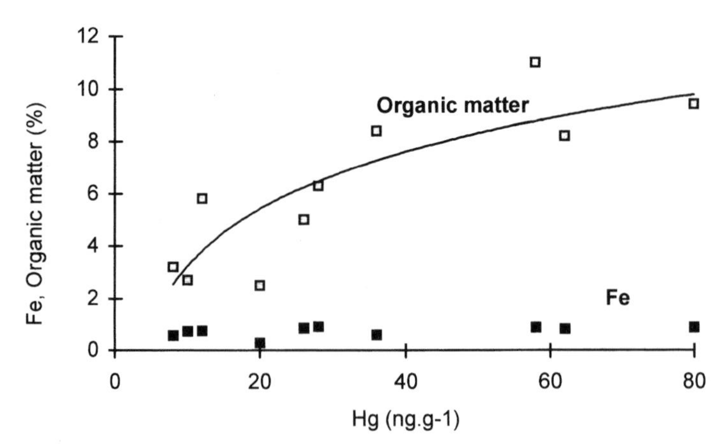

Fig. 4.6. Relationships between Hg content and the organic matter and iron concentrations of surface sediments of remote lakes in Pantanal, central Brazil

The relationship between Hg and organic matter, however, has not been detected in partitioning studies of white water river sediments receiving elemental Hg^0 directly from dredges (Pfeiffer et al. 1993), nor in rivers receiving elemental Hg^0 from riverbank mining (Brabo 1992). No correlation between organic matter and Hg has been found in suspended matter in such rivers either (Brabo 1992). Therefore, organic matter seems to be an important substrate for Hg only in sediments receiving indirect input of ionic Hg as Hg^{2+}, either from leaching of tailings or from the atmosphere.

On the other hand, in surface-active sediments from rivers draining tailings in central Brazil, significant positive correlations ($r = 0.72, n = 16$) were found between Hg and iron (Rodrigues 1994). The different geochemical partitioning of Hg in different environments is clear. While in active river sediments iron, probably as oxi-hydroxides, controls a significant fraction of Hg, in lake environments organic matter is probably the major substrate of Hg fixation in bottom sediments.

In the Madeira River sediments, Pfeiffer et al. (1993) found from 70 to 95% of the total Hg concentration present as elemental Hg^0. Mercury sulfides could represent from 1 to 21%, whereas Hg bound to fulvic and humic acids ranged from 1 to 16%. Soluble and exchangeable Hg accounted for only 4% of the total Hg present in sediments (Petrick 1993; Pfeiffer et al. 1993).

In white water river sediments Pfeiffer et al. (1993) and Petrick (1993) also found that more than 75% of the total Hg concentration was easily reducible. In creek sediments of the Poconé mining district, approximately 18% of Hg present was easily reducible by $SnCl_2$ (Tumpling et al. 1993b).

Most surveys on Hg concentrations in fluvial sediments of rivers affected by gold mining were done in the clay-silt fraction of sediments (< 63 μm). Few studies considered the effect of grain size upon the Hg distribution. Since metallic mercury is the major form of Hg released into these rivers, grain size seems to play a minor role in determining Hg sediment content. Figure 4.7a, b shows Hg distribution in river sediments of different granulometry in the Cumaru mining site, Pará State, based on the work of Ramos and Costa (1990). Along the river there are various points of Hg release, which explains the multiple peak distribution found (and also confirms the low mobility of metallic Hg released), in two different grain-size sediment fractions. It is clear that Hg distribution is independent of grain size. This is probably a result of the low Hg^{2+} concentration in river water ($< 0.02\ \mu g\ l^{-1}$), which would be preferentially adsorbed on fine particles. In the Mawddach River, North Wales, Fuge et al. (1992) also showed no preference of Hg for fine particles.

In some tailings, however, some workers found an enrichment of Hg in the fine fraction of freshwater sediments (Prokopovich 1984). Therefore, as

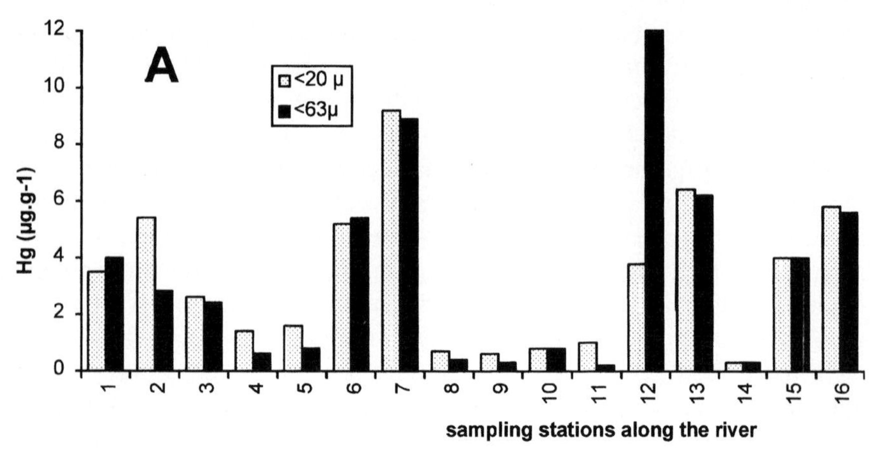

Fig. 4.7. Mercury concentrations in sediments of different grain size along the drainage of the Cumaru mining site, Para State, northern Amazon. (Ramos and Costa 1990)

with organic matter interactions, grain-size effects are only significant where indirect inputs of ionic Hg forms occur, but not where direct release of elemental mercury dominates.

4.3.3
Hydrological Effects on Hg Dispersion

The relationship between Hg dispersal and erosion and transport of suspended particles suggests that the dispersal mechanisms should vary according to the season, either through intensive erosion during the rainy season or through the effect of dilution on the existing Hg content in rivers. Few studies, however, measured Hg concentrations in environmental samples over a yearly cycle. The existing results are therefore still preliminary.

Rodrigues et al. (1992) measured total Hg concentrations (dissolved + particulate) in water of the Baía River, southern Brazil, and found that during the dry season, Hg concentrations ranged between 0.05 and 0.88 μg l^{-1}, while during the rainy season Hg concentrations ranged between 1.0 and 12.0 μg l^{-1}. Also, the variability among data during the same sampling period was much larger during the rainy season, suggesting pulse contamination of Hg as a consequence of leaching and transport of contaminated soil particles to the river.

Tumpling et al. (1993a) showed significant seasonal variation in the concentration of volatile Hg in waters draining tailings in Mato Grosso State, central Brazil, with concentrations ranging from 0.3 ng l^{-1} during the dry season to 5.2 ng l^{-1} during the rainy season, suggesting stronger leaching of Hg from tailings.

Based on these results and bottom sediment concentrations, they suggested that Hg-contaminated particles from tailings are transported during the rainy season and deposited in sediments of the drainage. After this, remobilization of contaminated particles from the sediment surfaces may take place, resulting in a decrease in Hg concentrations in the drainage bottom sediments and exportation to areas away from mining sites (Fig. 4.8).

In the Carajás mining site, Fernandes et al. (1990a, b, 1991) found contrary results. They measured the highest total Hg concentrations in river water (dissolved + particulate) during the dry period, ranging from 0.10 to 0.74 mg l^{-1} (when total suspended solids concentrations varied between 25 and 100 mg l^{-1}), while lower concentrations were found during the rainy season (< 0.05 to 0.38 mg l^{-1}, when total suspended solids concentrations varied from 50 to 400 mg l^{-1}). They explained their results as the effect of dilution and owing to the fact that mining activities in the area are restricted to the dry season. Table 4.2 compares the three studies. The difference in mining procedure seems to be a key factor controlling mercury dispersal. In areas where gold is mined from soil or rocks, during the rainy season the leaching and erosion of contaminated particles result in higher total Hg concentrations in rivers, while in areas where mercury is mined from river bottom sediments, Hg concentrations in water are lower during the rainy season owing to dilution and the decrease in the mining activity itself. At the Lerderderg River, Victoria, Australia, Bycroft et al. (1982) showed a significant decrease in Hg concentration in surface bottom sedi-

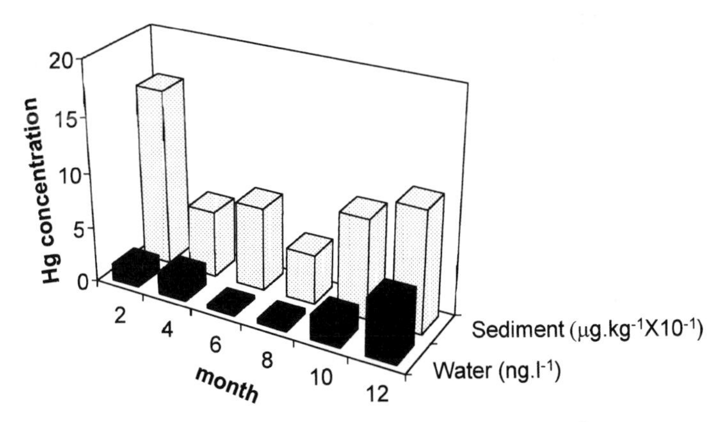

Fig. 4.8. Seasonal variation in dissolved metallic mercury (Hg^0) in water and total mercury in bottom sediments in a drainage of gold mining tailings in Poconé, central Brazil. (Adapted from Tumpling et al. 1993a, b)

Table 4.2. Seasonal variation in mercury total concentrations (dissolved + particulate) (in μg l^{-1}) in water from different mining sites

Study[a]	Dry season Hg	Rainy season Hg	Dry season Hg
A	0.41±0.21	2.95±3.26	0.31±0.17
B	0.32±0.18	0.11±0.26	–
C	0.30±0.10	5.20±1.00	–

[a] A, Paraná-Baía River (Rodriguez et al. 1992); B, Carajás Basin (Fernandez et al. 1990a,b, 1991); C, Poconé (Tumpling et al. 1993b), concentration in C is in ng l^{-1}, and refers to dissolved Hg only.

ments collected in winter compared with summer samples. Dilution with Hg-poor sediments may cause this difference and seems to be typical of tropical areas under humid climate (Lacerda et al. 1993). Season also affects the bulk quantities of Hg transported in the Waverly area of Nova Scotia, Canada; the transported amounts are smallest during low discharge, although Hg concentrations in the particulate matter are higher during this period of the year (Mudroch and Clair 1986).

However, no study has reported in detail on the speciation of Hg present in water draining tailings, making the interpretation of concentration changes difficult owing to changing water chemical properties. Also, no characterization of particle mineralogy or origin throughout the year, which could change Hg concentrations by changing proportions between suspended particles from differently contaminated areas, has been attempted. Therefore, these results have to be interpreted with care.

The relationship between total suspended solids and Hg concentrations is conflicting for two rivers receiving Hg from riverbank mining in Pará State, Brazilian Amazon. Suspended particles are basically the same in the two rivers, as shown from X-ray and IR analysis (Brabo 1992). While in the Marupá River a significant positive correlation was found, at the Crepori River no correlation was found between the two variables. The site of origin of the suspended matter in the two rivers may be a reasonable explanation, since tailings are located at different sites in the two basins (Brabo 1992).

4.3.4 Dispersion of Mercury in River Systems

Mercury lost to rivers as metallic Hg^{o}, the fate of which depends mostly on physical and gravimetric processes, is preferentially accumulated in bottom sediments and generally presents very low mobility, being transported associated with bed sediments along the bottom of the river, commonly referred to as the bed load. In the Madeira River, for example, the major tributary of the Amazon River, with a mean flow of 23 000 to 49 000 m^{-3} s^{-1} during the dry and rainy seasons, respectively, Hg concentrations in sediments close to operating dredges can reach values of up to 157 µg g^{-1} decreasing to background values of < 0.04 µg g^{-1}) a few kilometers downstream (Pfeiffer et al. 1989 a, b; Malm 1991). A similar pattern was reported for the Lerderderg River, Australia (Bycroft et al. 1982). In smaller rivers, Hg dispersion is even more restricted with background concentrations being found only a few hundred meters from the source (Andrade et al. 1988).

Physical processes are apparently responsible for transport and preferential accumulation conditions for metallic Hg^0 in specific locations. This was observed in a fluvial system in the USA (Thiboudeaux 1979) and in various Amazon rivers (Malm 1991). Pieces of Hg amalgam were observed in some places upstream from waterfalls. Two amalgams collected upstream from the Teotônio Waterfalls, in the Madeira River, weighed 56 and 137 mg, with a Hg content of 53 and 65 %, respectively (Malm et al. 1993).

The low mobility displayed by Hg is mostly due to its chemical form. Hg emitted to rivers is present as elemental mercury with a high density of 13.6 and low reactivity under natural surface conditions. Drops of metallic mercury have been frequently found in exposed river sediments during the dry season in the Madeira River. These drops amalgamate other metals present in river sediments. Amalgams found in the Madeira River basin presented up to 17 % Au, 0.3 % Cd, and 0.08 % Cu, i. e. orders of magnitude higher than the local background concentrations of these metals (Malm et al. 1993). This has also been reported for Hg mining areas in Monte Amiata, Italy, where Hg droplets are commonly found in the beds of streams running under abandoned Hg smelting plants (Bargagli 1995). Gold grains collected in drainages affected by last century gold mining in North Carolina, USA, presented a patchy distribution of Hg, ranging from 1.21 to 45.4 % in weight of Hg. Gold grains can act as a sink to Hg, decreasing its mobility (Callaham et al. 1994). Mercury found together with gold grains in river sediments has been reported for many other gold mining regions (Fuge et al. 1992).

In another study in the Madeira River, Martinelli et al. (1988 a) reported high (up to 1.04 µg g^{-1}) Hg concentrations in floating macrophytes from oxbow lakes 100 km away from the nearest Hg source. Müller (1993) studied the deposition of Hg in the lower section of the Tapajós River, SE Amazon, and found evidence of Hg transport over several hundred kilometers from mining sites associated with the suspended particles. This suggests long-range Hg transport associated with suspended particles.

Setting particles at the mouth of the Tapajós River still present high Hg concentrations (Fig. 4.9). Nriagu et al. (1992) suggested that nearly 70% of the mercury released into the Madeira River is exported downstream as far as the Amazon River, 1000 km downstream.

The dredging and consequently the resuspension of bottom sediments, which can multiply by 1000 the normal concentrations of total suspended solids, facilitates this process. For example, total suspended solids concentration increased from 4.0 to 54 mg l^{-1} in the Teles Pires River, Mato Grosso (Rodrigues 1994). This increased suspension load may be an efficient scavenger of incoming Hg.

In the Katun River, West Siberia, fine suspended particles (0.1 to 10 μm) transport 75% of the total yearly amount of Hg eroded from Hg-rich deposits upstream. Over 4.3 tons $year^{-1}$ are transported this way (Sukhenko et al. 1992). At the Shubenacadie River headwater lakes, the total Hg transported from tailings by creeks to the lakes is associated with the inorganic fraction of suspended particles, which show a Hg content ranging from 0.71 to 15.0 μg g^{-1} (Mudroch and Clair 1986).

In the Madeira River, suspended particle concentration can reach up to 1.25 g l^{-1} in the rainy season, dropping to nearly 0.1 g l^{-1} during the dry season (Ferreira et al. 1988; Martinelli et al. 1988b), with Hg concentrations ranging from 12 to 961 μg g^{-1} (Pfeiffer et al. 1993). Sediment transport and remobilization have also been claimed to explain fish contamination in the Madeira River collected circa 200 km downstream from major mining sites (Malm et al. 1990). Suspended particles have been demonstrated to be the main transport pathways of various trace metals in other Amazonian rivers (Gibbs 1973; DePaula 1989).

Brabo et al. (1991) found that pelitic bottom sediments are easily remobilized to the water column in South Amazonian rivers affected by gold

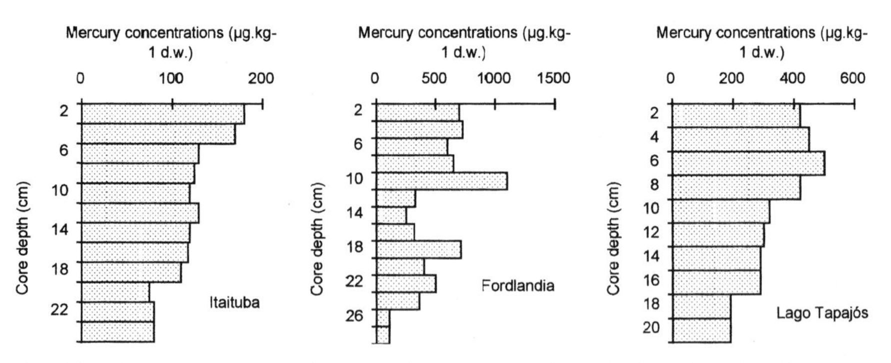

Fig. 4.9. Mercury concentrations in sediment cores from the Tapajós River between Itaituba and Santarém, southeastern Amazon

mining. This results in high Hg concentrations in suspended particles, ranging from 0.2 to 1.1 $\mu g\ g^{-1}$ dry wt., similar to the concentrations found in bottom sediments which ranged from 0.1 to 3.0 $\mu g\ g^{-1}$ dry wt. This mechanism is reported as the major process controlling mercury dispersion in these rivers.

Despite its obvious importance as an efficient Hg dispersion mechanism, few studies have analyzed the Hg content in TSS (total suspended solids) in gold mining areas. A good dataset is available for the Tapajós River and some of its tributaries, SE Amazon. Mercury concentrations in suspended particles close to mining sites are very high, up to 8.0 $\mu g\ g^{-1}$. However, relatively high concentrations (0.5 $\mu g\ g^{-1}$) are still found hundreds of kilometers away from mining sites. Setting particles at the mouth of the Tapajós River still present high Hg concentrations (Müller 1993). This suggests the efficiency of the long-range transport of Hg associated with suspended particles (Brabo et al. 1991; GEDEBAM 1991; Thornton et al. 1991).

Another example of long-range transport of Hg from gold mining areas was demonstrated for the Paraíba do Sul River, southeastern Brazil. Gold production in this river is small (between 20 and 50 kg year^{-1} only). However, gold mining is the major source of Hg to the lower basin of the river. It is estimated that from 150 to 300 kg of Hg has been released into the river since 1985 (Lacerda et al. 1993a).

Figure 4.10 shows the distribution of Hg in river, estuary, and continental shelf sediments of the Paraíba do Sul River. Concentrations drop

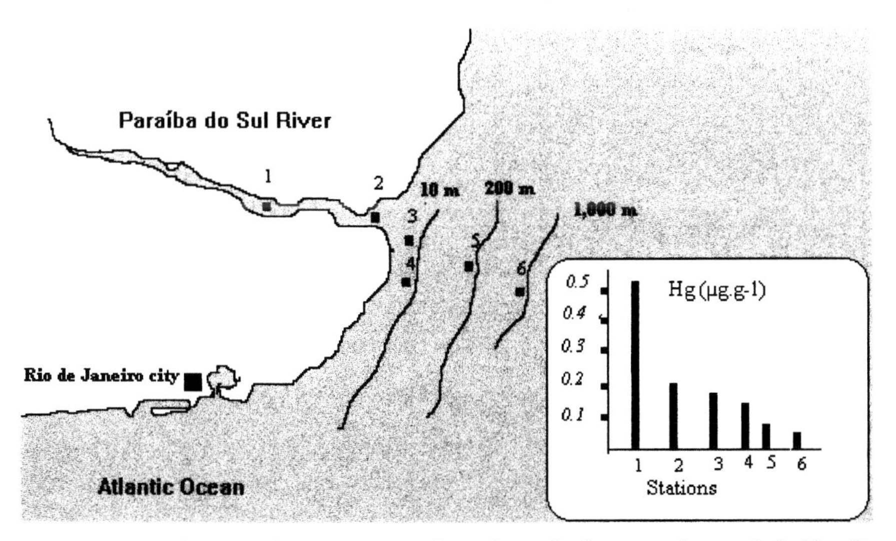

Fig. 4.10. Distribution of Hg concentrations through the estuarine and shelf sediments off the Paraíba do Sul River, southeastern Brazil. (After Lacerda et al. 1993a)

from a maximum of 550 μg g^{-1} in the fluvial sediments to 110 to 150 μg g^{-1} in the estuary and river plume. Only at a water depth of over 70 m, nearly 70 km offshore, did Hg concentrations reach background values, between 30 and 50 ng kg^{-1} (Lacerda et al. 1992). The importance of the transport of Hg associated with suspended particles through the coastal zone to deep waters has also been reported for other marine areas, such as the Mediterranean (Cossa and Martin 1991) and the Adriatic Sea (Ferrara and Maserti 1992).

4.4 Tropical Reservoirs

One major issue of Hg dispersion from gold mining areas is the impact on ecosystems not directly related to and even far from any mining sites. The large atmospheric component of Hg dispersion from gold mining will enlarge the potential impacted area to a regional rather than only a local scale. One of the most impressive examples of this dispersion phenomenon is the increase in Hg concentrations in artificial reservoirs, even when these water bodies are located far from the emission sites (Jackson 1988). This phenomenon is partially due to the large fraction of Hg dispersed through the atmospheric pathway and to the complex biogeochemical conditions found in recently impounded reservoirs (Jackson 1986).

Tropical regions have one of the largest hydroelectric power potentials in the world. Presently, large dams are under construction or already in operation in many tropical countries. For example, in the Brazilian Amazon alone, the area flooded by hydroelectric power plant dams reaches 4800 km^2 and is expected to double by the beginning of the next century (Fearnside 1994). All future scenarios forecast more dams to be completed by the beginning of the next century.

Artificial reservoirs create completely different hydrological and hydrochemical settings in their basins, mostly as a result of changing river systems into huge, generally shallow and nearly stagnant lakes. Major hydrochemical and hydrobiological changes involve nutrient cycling, primary production, and sedimentation (Esteves 1988).

The building of reservoirs in the tropics is generally achieved by inundating large areas of primary forest. The decaying biomass results in high rates of oxygen consumption during the microbial decomposition of organic matter. Large amounts of nutrients, in particular nitrogen compounds, are also liberated into the water during this process. Thermal stratification produces an anoxic, nutrient-rich hypolimnion. Breaking of the thermocline in winter provides the phytoplankton in the open waters and the macrophytes along the margins with an unlimited supply of nutrients, which, coupled with high temperatures and intense insulation, results in very high primary production. During these phytoplankton blooms and macrophyte outbreaks, virtually all dissolved chemicals are incorporated into organisms.

The quiet environment of most reservoirs allows the accumulation of fine (clay and silt) particles in bottom sediments. The typical reservoir sediment is generally anoxic and rich in organic matter, and with a metabolism based on sulfate reduction or methanogenesis. The microorganisms involved in these reactions play a key role in the alkylation of trace metals (Jackson 1986; Esteves 1988).

4.4.1 Atmospheric Inputs

In the process of gold extraction, mercury is lost to the atmosphere as Hg vapor (Hg^o). In the atmosphere, a portion of it may adsorb to particles and return to nearby areas. An important fraction, however, is oxidized to Hg^{2+} through reactions mediated by ozone, solar energy, and water vapor. Once formed, ionic mercury (Hg^{2+}) is removed from the atmosphere by rain and is deposited in terrestrial and aquatic environments, where it may undergo further reactions, including organification and assimilation by the biota. These mechanisms will provide a long-range dispersion of mercury, capable of reaching areas far from the emission sources.

The atmospheric inputs of Hg in mining areas are variable, depending on the scale of the production process. However, bulk deposition can range from near-background values of about 30 $\mu g\ m^{-2}\ year^{-1}$ to very high values of up to 120 $\mu g\ m^{-2}\ year^{-1}$ (Lacerda et al. 1992). For the Amazon region, for example, Hg inputs to the atmosphere reach 32 $g\ km^{-2}$. Therefore, impounded areas of hydroelectric dams in this region may receive up to 160 kg Hg $year^{-1}$ directly from the atmosphere. This significant input is supposed to be oxidized Hg forms and Hg associated with particles. Both forms are capable of being methylated by aquatic microbiota (Guimarães et al. 1994b).

Apart from direct atmospheric deposition over the reservoir itself, deposition over the large watersheds, typical of tropical reservoirs, may eventually contribute to incoming fluvial systems.

4.4.2 Fluvial Inputs

Mercury lost to rivers as metallic Hg is preferentially accumulated in bottom sediments and generally presents very low mobility. Concentrations in bottom sediments show a fast decrease from high values close to releasing points to background values a few kilometers downstream. However, studies on Hg partitioning between dissolved and particulate phases (Nriagu et al. 1992) and on Hg transport associated with particles (Lacerda et al. 1991b) suggest that long-range Hg transport associated with suspended particles occurs in most river systems. The dredging and resuspension of bottom sediments by mining operations facilitate this process. The load of suspended particles in tropical rivers can reach up to 1.0 g l^{-1} and has recently increased due to generalized deforestation in most tropical countries. These particles can transport a fairly high amount of Hg, and various studies have demonstrated that they are the major transport pathway of trace metals in tropical rivers (Gibbs 1967, 1973, 1976).

Other results from gold mining tailings show that Hg is transported in association with suspended particles after erosion of tailings by rain. Mercury concentrations in suspended particles increase from one to two orders of magnitude during rains, and this may also provide long-range transport of Hg through fluvial systems (Lacerda et al. 1991b), at least on a seasonal basis (Tumpling et al. 1995).

Once reaching artificial reservoirs, suspended particles will settle due to the long residence time of water. Also, during deposition, particles will be submitted to completely different hydrochemistries and alterations of the Hg species. Furthermore, reactions with autochthonous material is expected.

4.4.3 Mercury Cycling in Artificial Tropical Reservoirs

At least two major autocthonous inputs of Hg to tropical reservoirs can therefore be established. The atmospheric input which, depending on the location of the reservoir, can be the major Hg source; and the riverine input where Hg is associated with the incoming suspended matter. During flooding of soils and vegetation, either during the formation of the reservoir or during the rainy season, a significant amount of Hg may be released from soils and forest litter upon flooding. For example, the Hg content of recently flooded litter in a French Guyana reservoir showed concentrations of about 100 μg kg^{-1}. Degradation products such as organic-rich muds and fine debris were enriched with Hg, with concentrations ranging from 146 to 177 μg kg^{-1}. This suggests that bacterial degradation of organic matter at the sediment–water interface results in an enrichment of fine suspended particles through the release of Hg contained in decaying plant material (Roulet and Lucotte 1994). This autochtonous source of Hg may be significant, depending on the soil and litter Hg concentrations.

A simplified model for Hg cycling in tropical reservoirs is presented in Fig. 4.11. Atmospheric Hg reaching the reservoir will probably be Hg^{2+}. This species is highly reactive with dissolved ions present in the reservoir waters. During stratification periods with low primary productivity, Hg^{2+} adsorbs onto incoming suspended particles and is scavenged to the bottom. A significant fraction, however, may remain in solution forming complexes with inorganic species (e.g., Cl^-), and most probably with soluble organic ligands (Fig. 4.11). Dissolved organic matter is a key factor in increasing Hg solubility in aquatic systems (Meili 1991b, c; Forsber et al. 1994).

Without stratification or after destratification, a general anoxia or sub-anoxia may dominate the reservoir water column. Under such situation

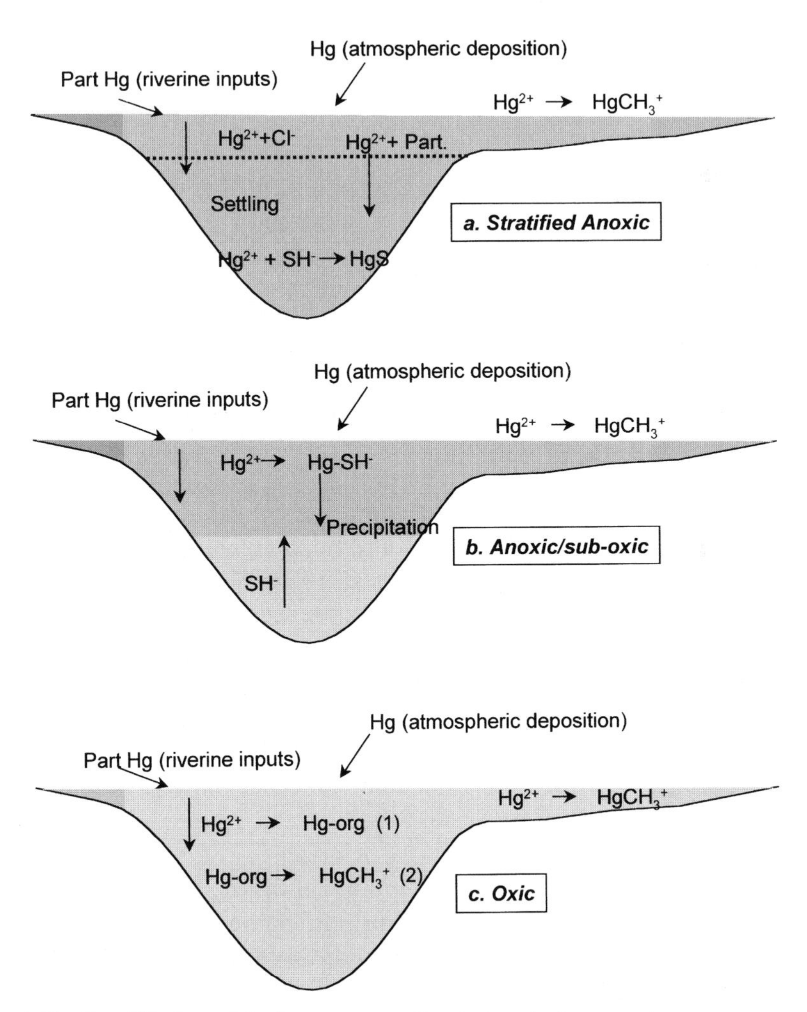

Fig. 4.11. Simplified model of mercury cycling in tropical reservoirs

nearly all Hg^{2+} will precipitate as highly stable sulfides (SH^-) which accumulate in the anoxic hypolimnion due to bacterial sulfate reduction. Precipitated Hg sulfides may be permanently buried in the reservoir sediments due to the relatively high sedimentation rates (Fig. 4.11).

When oxygen levels rise, phytoplankton blooms may occur due to the high availability of bottom-derived nutrients. Here, Hg^{2+} will eventually be incorporated in the phytoplankton cells [Fig. 4.11, Eq. (1)] and released as soluble Hg-organic complexes upon microbial decomposition, either in the epilimnion or in the anoxic hypolimnion. High rates of Hg methylation may then occur, since favorable conditions, including slight acidity, high organic substrate concentrations, and significant microbial activity will dominate the water column and the surface of sediments [Fig. 4.11, Eq. (2)]. High rates of Hg-methylation have been measured in large tropical reservoirs (Guimarães et al. 1994).

Methyl-Hg is highly soluble and stable. Therefore it will move up the water column after destratification. In the epilimnion, methyl-Hg will be rapidly incorporated in fish and may also be transported downstream dissolved in reservoir water.

Another important pathway resulting in Hg methylation may involve incoming suspended particles. Mercury-enriched suspended particles tend to concentrate near incoming rivers and deposit in shallow areas of the reservoir where macrophytes dominate. In such areas, Hg deposition can reach up to 240 μg m^{-2} $month^{-1}$ (Aula et al. 1994). These shallow areas are characterized by intense microbial activity, due to the permanent input of fresh, oxidizable organic matter from the macrophytes themselves.

Dissolved organic matter (DOM) significantly enhances the solubility of inorganic and organic Hg species and decreases the sorption of organic Hg onto sediments (Melamedi et al. 1995). The effects of increasing concentrations of DOM upon Hg species are shown in Fig. 4.12. As a result, in the presence of DOM, exceptional conditions for the formation and permanence in solution of methyl-Hg exist in these marginal areas of reservoirs. Elevated temperatures and acidity due to the release of organic acids from the vegetation will also contribute significantly to these processes. In fact, floating mats of freshwater macrophytes in artificial reservoirs have shown elevated methylation rates (Guimarães et al. 1994b).

Apart from having ideal biogeochemical conditions for Hg methylation, the macrophyte-dominated littoral of an artificial reservoir also presents a high fish biomass year-round. In these areas therefore Hg methylation will occur very rapidly and methyl-Hg will be immediately incorporated in the fish biomass.

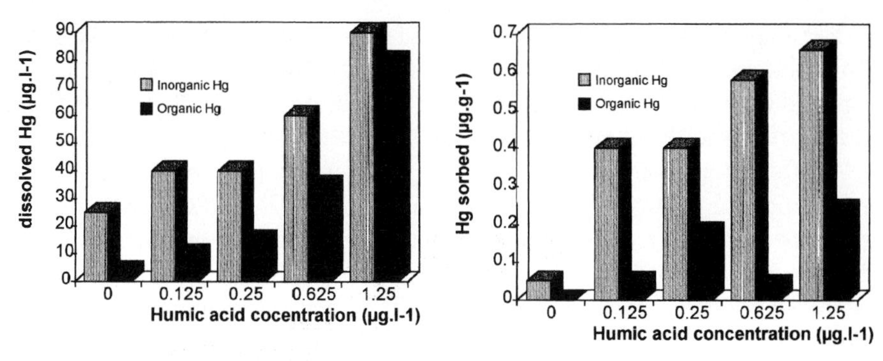

Fig. 4.12. The effects of dissolved organic matter on the solubility and sorption capacity of mercury species. (After Melamedi et al. 1994)

The only complete survey of Hg distribution in tropical artificial reservoirs confirms the model described above. The Tucuruí reservoir, one of the largest Amazonian reservoirs, was surveyed by Aula et al. (1994). Table 4.3 summarizes their results.

Mercury concentrations in the abiotic compartments of the reservoir are relatively low. In fact, in many places, they do not differ from background concentrations measured by other authors in geographically similar areas (Lacerda et al. 1991b). The same is true for the concentrations found in low trophic level organisms, such as aquatic macrophytes, snails and herbivorous mammals. However, Hg concentrations in carnivorous fish, caiman meat and human hair from local fishermen were extremely high, confirming the large availability of methyl-Hg in the reservoir. The consequences of such high Hg concentrations in biotic compartments will be discussed in detail in the next sections.

Table 4.3. Mercury distribution in biotic and abiotic compartments of the Tucuruí Reservoir, SE Amazon. (According to Aula et al. 1994)

Compartment	Range of Hg concentrations ($\mu g\ kg^{-1}$)	Average (when available) ($\mu g\ kg^{-1}$)
Suspended particles	59–300	–
Bottom sediments	37–130	–
Aquatic macrophytes	30–120	–
Aquatic invertebrates	11–170	57
Aquatic mammals	7–26	15
Caimans	1700–3600	1900
Fish	60–2600	–
Human hair	6–241	65

The major stock of Hg in the Tucuruí reservoir is the flooded topsoil, which has over 8.8 tons of Hg; the bottom sediments have 3.5 tons. Inundated litter and aquatic macrophytes contribute to only a small fraction of the total amount of Hg in the reservoir, 0.08 and 0.03 tons, respectively.

Artificial reservoirs in the tropics act as reactor vessels for Hg, receiving inorganic and particulate Hg and cycling it through pathways which will generate methyl-Hg. However, a significant proportion of the incoming Hg may be permanently buried in the reservoir sediments as Hg sulfides. Methyl-Hg formed in the reservoir may be either incorporated in the fish biomass or exported in solution downstream. In the first case, it will seriously affect the utilization of the reservoir for fish culture, in the second, it may undergo long-range transport and affect areas far from the original Hg source or negatively affect agriculture if reservoir water is used for irrigation.

The many possible interactions between Hg generated in gold mining and under reservoir conditions may result in a decrease in the economic value of the many alternative uses of reservoir waters. However, intensive research is needed to test the various hypothetical pathways of Hg cycling described here. This research is urgent since gold mining operations will be increasing in the future due to the economic constraints of many tropical countries which, on the other hand, may also witness a boom in reservoir construction imperative for the generation of the energy necessary for the region's development.

4.5 Mercury Methylation

In the humid tropics where most present-day gold mining occurs, abundant organic matter, shifting anoxic to oxic conditions and high temperatures, is typical of most floodplain areas, including lakes and forest rivers. Under such conditions, Hg methylation rates are probably very high (Lacerda et al. 1989). Therefore, the formation of methyl-Hg surely plays a key role in further contamination of organisms including man.

The solubility and stability of methyl-Hg in natural waters make the methylation process a significant dispersion mechanism of Hg in freshwater systems (Melanetti et al. 1995). Figure 4.13 shows the effect of methylcobalamine on the adsorption of Hg^{2+} onto sediments from a tailings drainage in central Brazil.

The formation of methyl-mercury (MeHg) under natural conditions, in a process mediated by microbial activity, was first demonstrated Jensen and Jernelov (1969) in sediments of a Swedish lake. This offered a possible explanation for the so far intriguing evidence that most Hg in fish and

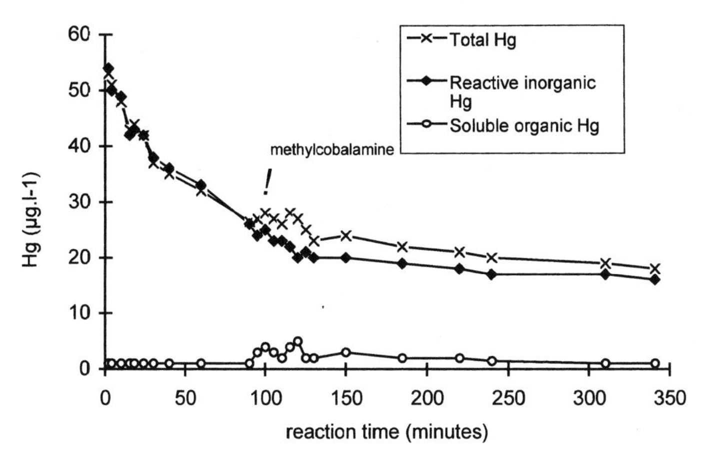

Fig. 4.13. The effect of methylcobalamine on Hg adsorption onto sediments of a tailings drainage in central Brazil. (After Melaneti et al. 1995)

other aquatic organisms is present as MeHg, while Hg in industrial effluents and fallout is mainly inorganic. A great variety of microorganisms are able to methylate Hg, including strepto- and staphylococci, lactobacilli, yeasts, and different fungal cultures. Mercury can be methylated by aerobic and anaerobic bacteria but only a few obligate anaerobes can form MeHg. Additionally, the latter produce less MeHg than facultative anaerobes (Robinson and Tuovinen 1984). Hg-methylating microorganisms occur in most environmental substrates and even in fish and human intestinal contents (Furutani et al. 1980).

The mechanism of Hg methylation is still not well understood, despite its evident toxicological relevance. It appears to involve methylcobalamine (vitamin B12) as the donor of methyl groups to Hg^{2+} in which a carbanion transfer from the methylcobalamine to mercurial species in ionic or weakly complexed forms takes place (Landner 1971; Kersten 1988). Mercuric Hg^{2+} is believed to be the direct precursor of MeHg, though the formation of dimethyl-mercury from phenyl-mercury acetate by some microorganisms in soil and water has already been reported (Matsumura et al. 1971). Mercury can be methylated by abiotic or photochemical processes but their significance is relevant only in limited areas.

The dominance of the biological pathway in methyl-Hg production imposes temporal fluctuations in the availability of methyl-Hg for natural processes of transport and degradation. In central Asian provinces of Russia, 20 % of Hg migrates as methyl-Hg in natural waters, in particular during the period of maximum biological production (Ermanov 1995). Changes in biological production are also responsible for the changing atmospheric Hg inputs to Antarctica during the last 40 000 years (Vandal et al. 1993).

Once formed, MeHg is able to resist many environmental stresses except specific biochemical degradation processes, again mediated by microbial activity. These demethylation mechanisms are much better understood than methylation itself. Bacterial resistance to organic and inorganic Hg compounds is determined by plasmids that often also encode the resistance to other heavy metals and to antibiotics (Robinson and Tuovinen 1984). The resistance to MeHg involves the detoxification of this compound by cleavage of the C–Hg bond by the organomercurial lyase enzyme followed by the elimination of the metal from the medium by conversion to the volatile Hg^0 through the action of a mercury reductase enzyme, in such a way that mercury resistance and volatilization are essentially synonymous. The production of both enzymes is induced by exposure to Hg. Resistance to this metal was shown to be governed by at least four genes whose genetic and physical structure has already been mapped (Robinson and Tuovinen 1984), leading some authors to suggest that native bacteria from contaminated sites could be genetically manipulated to enhance their ability to demethylate and volatilize MeHg (Winfrey and Rudd 1990). Abiotic processes can also reduce Hg^{2+} ions as shown by Alberts et al. (1974) who found that humic acids could stimulate Hg^0 evolution.

Usually, due to the relatively complex procedures involved (Ramlal et al. 1985, 1986), only net methylation, the balance between MeHg production and decomposition, is measured. These studies have shown, however, that methylation and demethylation react in a complex and variable way to parameters such as Hg^{2+} and MeHg concentration and availability, pH, redox potential, composition of the microbial population, and other synergistic or antagonistic effects of environmental and biological processes (Korthals and Winfrey 1987).The differences in the optimal conditions for these activities may render them mutually exclusive when present in the same organism and may define microenvironments in which one activity predominates (Robinson and Tuovinen 1984).

In lakes, MeHg formation has long been considered the main source of MeHg to fish but Hg methylation in soils has already been demonstrated (Rogers 1976, 1977), and MeHg was measured in run-off and precipitation (Lindqvist et al. 1984; Lee and Hultberg 1990). These findings, together with the observation of high Hg concentrations in fish from colored lakes and the correlation between DOC and MeHg in humic waters, suggest that the terrestrial ecosystems may be relevant MeHg contributors to lakes and rivers, possibly due to abiotic methylation by fulvic and humic acids in the watersheds (Lee et al. 1985). The relative importance of terrestrial and in-lake MeHg production would therefore vary with factors such as the drainage area and the residence time of the lake water (Winfrey and Rudd 1990).

Fish in reservoirs and artificial lakes tend to present marked increases in Hg content after impoundment and it may take many years to return to the pre-impoundment Hg concentrations (Gilmour and Henry 1991). It has been hypothesized that the flooding of large (and generally forested) areas may cause a sudden mobilization of Hg present in soils and vegetation, the methylation of which will be favored by the increased microbiological activity and reducing conditions associated with the decomposition of the flooded vegetation (Jackson 1986, 1988). Recently, average total Hg concentrations of up to 2.8 $\mu g\ g^{-1}$ were measured on carnivorous fish samples (piranha, *Serrasalmus* sp.) of the Tucuruí reservoir on the Tocantins River, eastern Amazon (Aula et al. 1994).

Differences in the net methylation rate controlling factors among environments complicate the construction of general predictive models on Hg methylation and bioaccumulation that would apply to a wide variety of environments (Winfrey and Rudd 1990). Most research on Hg behavior has taken place in temperate or boreal aquatic systems and very little is known about Hg biogeochemistry in tropical waters. In the particular case of the Amazon, the diversity and complexity of the aquatic ecosystems and the widely dispersed and largely unpredictable nature of gold mining activities are additional difficulties.

The data available to date on MeHg concentrations in the environment and in human populations in the Amazon are scarce but point to some trends that are already familiar (Malm et al. 1990). Malm et al. (1994) showed that Hg in fish of the Madeira and Tapajós Rivers as well as in the hair of fish-eating dwellers in these areas was essentially MeHg.

An important fraction of our present knowledge of Hg methylation and demethylation has been obtained through radiochemical techniques. The advantages and limitations of this approach have been reviewed in detail by Furutani and Rudd (1980) and Gilmour and Henry (1991). Incubations of environmental samples with ^{203}Hg followed by extraction and measurement of $Me^{203}Hg$ formed during incubation are especially useful to test hypotheses on the influence of single variables on Hg methylation rates and on preferential methylation sites. However, the obtained rates are considered estimates of maximum potential Hg methylation rates since the carrier Hg added along with ^{203}Hg increases Hg concentration and availability in the sample.

Simplified versions of the radiochemical technique originally described by Furutani and Rudd (1980) have been developed and applied by Guimarães (1992) and Guimarães et al. (1994a, b) to the determination of net Hg methylation rates in sediment, water and soil samples of the Madeira and Tapajós River basins, Brazilian Amazon. Methylation was not detectable in unfiltered water samples of these rivers incubated with $^{203}HgCl_2$ at Hg concentrations equivalent to 5–20 μg l^{-1} for up to 3 days, nor in forest soil–water suspension samples incubated under the same conditions. High methylation rates were, however, found in surface sediment samples of the Madeira River area, as shown in Table 4.4. Low rates of 10^{-3} and 10^{-5}% $g^{-1} h^{-1}$ were found for the samples of the typical white-water Madeira River, where low methylation rates were anticipated due to its high pH, conductivity and suspended sediment load. In the Mutum-Paraná River and Novo creek, typical black water forest streams, the rates reached 10^{-2}, which is already comparable or even higher than seasonal methylation peaks observed in temperate lakes (Furutani and Rudd 1980; Ramlal et al. 1986; Korthals and Winfrey 1987). Rates higher than this by one to two orders of magnitude were, however, found in the highly organic flocculent surface sediment of the flooded forest area of the Samuel reservoir and in similar sediments collected just downstream of the dam, both sites characterized by mildly reducing conditions. These data suggest that intrinsic methylation rates are high in the Samuel reservoir, which may explain the finding of total Hg concentrations approaching or exceed-

Table 4.4. Potential net ^{203}Hg methylation rates in some surface sediment samples from the Amazon region, Brazil. Average and range for samples of water-sediment slurries of up to 100 ml incubated in situ for 15–24 h with 44 kBq of ^{203}Hg (approx. 2 μg Hg). (Guimarães 1992; Guimarães et al. 1993 a, b). Incubation of samples kept at 4 °C for 69 days

Sampling site	Methylation rate (%.$g^{-1} h^{-1}$)
Madeira River, upstream of Mutum-Parana River	6.2 (5.1–7.3) × 10^{-5}
Madeira River, upstream of Teotonio Falls	1.5 (1.4–1.6) × 10^{-3}
Mutum-Parana River	1.0 (0.8–1.2) × 10^{-2}
Novo Creek, midstream	3.0 (2.8–3.2) × 10^{-2}
Ponds in tailings of abandoned mining areas	8.5 (7.8–9.3) × 10^{-3}
Forest rivers downstream of mining sites	6.1 (3.9–2.1) × 10^{-3}
Forest rivers in pristine areas	3.3 (2.8–3.8) × 10^{-2}
Rato River (mining site)	6.5 (3.5–9.4) × 10^{-2}
Tapajós River	1.2 (0.7–1.6) × 10^{-2}
Jamari River, approx. 500 m downstream of Samuel Reservoir	6.6 (3.2–9.9) × 10^{-1}
Samuel Reservoir, flooded forest	6.9 (6.4–7.4) × 10^{-1}
Floodplain lakes, central Amazon	1.8 (0.2–3.7) × 10^{-1}
Floating macophyte beds, central Amazon	1.6 (0.5–2.8) × 10^{-1}
Samuel Reservoir, approx. 100 m from the dam	2.9 (2.1–3.7) × 10^{-3}

ing the 0.5 μg g^{-1} safety limit in many carnivorous fish caught in the reservoir, despite the very low Hg concentrations in the bottom sediment (Table 4.4).

Marginal floodplain lakes of the Rio Negro basin also presented high methylation rates, ranging from 9.5×10^{-2} to 37×10^{-1}% $g^{-1} h^{-1}$. High rates were also obtained when macrophyte mats were incubated, reaching values as high as 2.8×10^{-1}% $g^{-1} h^{-1}$ (Guimarães et al. 1994b).

Petrick (1993) showed that approximately 80 % of total Hg in sediments of the Madeira River is in the metallic form. Since Hg^{2+} is believed to be the substrate for methylation, Hg oxidation will be a rate-limiting factor for Hg methylation. Gold mining in the Madeira River is performed by mechanical dredges and their operation favors Hg^0 oxidation by continually bringing huge amounts of sediment to the surface. In the Tapajós River and tributaries gold is extracted from the bottom sediment by small rafts but most frequently co-alluvial gold is extracted from soils of the riverbanks or watersheds. In the Tapajós River, sediments are not yet available, but methylation experiments similar to those described above for the Madeira River area indicated that Hg finds favorable conditions for methylation since it is released in gold mining areas down to the Tapajós River itself. Net methylation rates in the sediments of ponds in recently abandoned gold mining fields were in the 10^{-3} range, while in the sediments of streams that drain these mining areas and of the Tapajós River rates were one order of magnitude higher (Guimarães et al. 1993 a, b). No methylation was found in non-flooded soils.

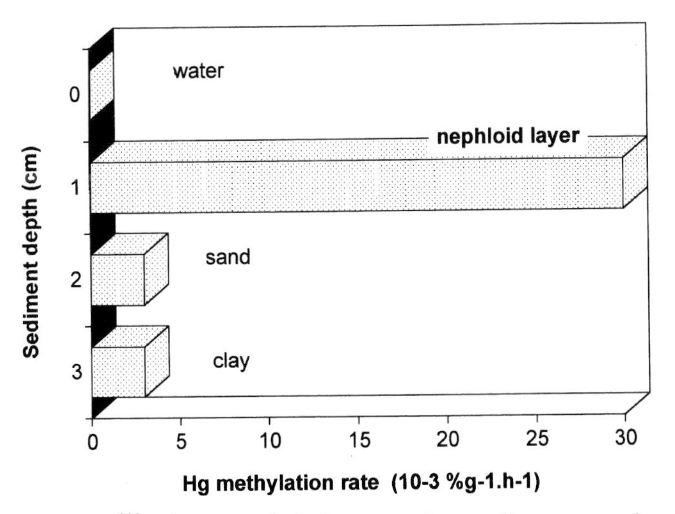

Fig. 4.14. Mercury (^{203}Hg) net methylation rates in a sediment core from Samuel Reservoir, Rondônia, SW Amazon. (After Guimarães et al. 1993b)

The vertical distribution of methylation rates is presented in Fig. 4.14. Methylation was not detected in the water column. Significant rates were only measured in the highly organic nephloid layer at the sediment water interface, nearly 30 times higher than the rates measured at deeper sediment horizons. No significant difference was found between methylation rates in clay and sand substrates (Guimarães et al. 1993b).

The data described strongly suggest that Hg methylation depends on the presence of easily decomposing organic matter as a source of energy to microorganisms in the sediment-water or pore-water interface. The nephloid layers in tropical lakes and reservoirs, and the interstitial waters of floating macrophyte beds, are basically an organic ooze, with an extremely high bacterial metabolism. Therefore these sites are ideal for Hg methylation to occur at high rates. This may result in extremely serious environmental problems, since floodplain areas are the most productive of aquatic systems in the tropics (Ayres 1994).

Unfortunately, only one survey has been completed on the concentration of this bioavailable form of mercury in water and sediments of Amazonian rivers (Padberg 1990), although others have been recently started in the region. The study of Padberg (1990) was carried on in the Tapajós River, where one of the largest mining sites in Amazon can be found. Table 4.5 summarizes the major results.

It is clear that methylation is presently occurring in this river system; despite the low concentrations found. Methyl-mercury concentrations reached nearly 10% of the total mercury content in water, but only 2% of the total mercury content of sediments.

Padberg (1990) found methyl-Hg concentrations in the Tapajós River ranging from 0.2 to 0.6 ng l^{-1}, representing from 3.8 to 16% of the total Hg concentration. The percentage of methyl-Hg found, although relatively high, is probably derived from leaching of methyl-Hg from marginal lakes and floodplains, since Guimarães et al. (1993a, b) found no measurable methylation rates in the water column of large rivers in the Amazon. This mechanism of methyl-Hg transport from marginal areas was first hypothesized by Lacerda et al. (1989) to explain the high Hg concentrations

Table 4.5. Total mercury and methyl-mercury concentrations in sediments of the Tapajós River, Pará State, northern Amazon. (After Padberg 1990)

	Total Hg ($\mu g\ kg^{-1}$)	Methyl-Hg ($\mu g\ kg^{-1}$)	Percent methyl-Hg
Itaituba	144	0.8	0.6
South of Itaituba	2.9–9.3	0.07–1.9	2.2

found in fish from large white waters rivers of the Amazon basin, which presents unfavorable conditions for Hg methylation and receives only non-reactive elemental Hg^{o} from dredges. Atmospheric deposition of Hg^{+2} rather than dissolved Hg^{o} or particulate Hg in rivers would supply the substrate for the microbially mediated methylation in marginal lakes and floodplain areas (Mason and Morel 1993).

5 Mercury in Biota

5.1
Terrestrial Vegetation

Terrestrial vegetation interacts with both soil and the atmosphere in the acquisition of Hg (Siegel et al. 1987). We have shown in previous chapters the importance of atmospheric Hg in the geochemical cycles in gold mining areas. Terrestrial plants can take up Hg from the atmosphere directly through gas exchange at the stomatal level or through membrane exchange with wet and dry precipitation (Browne and Fang 1978). These processes are likely to be enhanced in tropical vegetation due to the efficient mechanisms of chemical element acquisition from the atmosphere mediated by epiphyllous organisms, in particular algae and lichens (Jordan et al. 1980) at the crown leaf level, and those mediated by the high diversity and biomass of ephiphytes typical of most tropical ecosystems (Benzing 1981; Nadkarni 1984).

In most tropical regions where gold mining occurs, relatively pristine tropical forests dominate. The biomass of such ecosystems is particularly large. Therefore, depending on the Hg concentrations in plant tissues, the forest biomass may represent an important sink for Hg, as it certainly does for many other trace elements (Golley et al. 1975).

Recently, increasing deforestation rates of tropical forests has increased preoccupation with the potential release of Hg through forest burning in the tropics, resulting in the first estimates of Hg stored in the vegetation (Veiga et al. 1994; Lacerda 1995). These estimates, however, showed that forest burning is a very small source of Hg in tropical areas (Lacerda 1995). Notwithstanding the recognition of forest burning as a very minor source of Hg to the tropics, the first estimates of Hg content in the forest biomass were obtained. Lucotte (1994) and Roulet and Lucotte (1994) presented one of the first inventories of Hg concentrations in tropical forests. Surveying forests in French Guyana and northern Brazil, they found relatively high Hg concentrations in forest leaves of around 0.10 $\mu g\ g^{-1}$ (0.02–0.29 $\mu g\ g^{-1}$). This compartment, however, represents less than 5% of the total biomass. The larger fraction, making up over 80%, consists of trunks and branches (Golley et al. 1975),which present very low Hg concentrations of approximately 0.01 μg g^{-1} only (Lucotte 1994). Estimates of Hg content stored in the plant biomass of tropical forests are therefore small, reaching only 3.0 to 5.0 g ha^{-1}. For example, using this estimate, the total Hg content of the entire Amazon forest biomass would reach only 2000 to 5000 tons (Lacerda 1995).

Although still based on very few surveys, Hg concentrations in natural terrestrial vegetation seem to be very low. Exceptions, however, may occur in the vegetation growing on tailings. Here, natural selection may result in metal-tolerant forms, capable of accumulating relatively large amounts of Hg (Ernst 1988). Mercury concentrations in higher plants from the Hg-Sb provinces of the northern slopes of Turkestan, central Asia, range from 0.1–184 μg g^{-1} (Ermenov 1995).

Some studies have been carried out on the vegetation cover of old mining tailings. At the Oldham tailings, Nova Scotia, vegetation coverage exceeds 100%, with species richness ranging from 12–15 species m^{-2}, which is higher than that found in abandoned pastures in the same region (Lane et al. 1988). Many of the plant species recorded in the area are metal-tolerant species or metal-tolerant genotypes derived from local populations. All species analyzed in these tailings presented abnormally high Hg concentrations, ranging from 0.2–3.5 μg g^{-1} (Lane et al. 1988). All terrestrial plant species studied presented higher Hg concentrations in roots.

Siegel et al. (1985), studying the vegetation on tailings in British Columbia, found that different concentration factors (enrichment factors) among species only occurred at low Hg concentrations in soils. When concentrations increased over 0.05 μg g^{-1}, enrichment factors were relatively constant among taxa, suggesting that they are metal-tolerant genotypes. Although only a few studies were done in Hg-contaminated tailings, it seems that the behavior of the vegetation is similar to any other vegetation growing on metal-rich tailings.

Much more data, however, are needed for a better comprehension of the fate of Hg in terrestrial vegetation affected by gold mining.

5.2 Aquatic Macrophytes

Aquatic macrophytes are more likely to accumulate Hg than terrestrial plants. The availability of Hg to these plants will be a function of the dissolved Hg in water. Under the more acid freshwater conditions typical of tropical systems, Hg availability should be high, and certainly much higher than in the soil or sediment solution.

Different habitats of aquatic macrophytes may result in different Hg concentrations, since the roots may effectively avoid Hg uptake from soils and sediments. Floating and submerged species, with roots in the water column, may absorb Hg also directly from the water through leaves and shoots. This has been shown both between and within plant genera (Kelly 1988).

Martinelli et al. (1988) compared Hg levels in floating and emergent macrophytes from a marginal lake in the Madeira River basin and found Hg concentrations of up to three orders of magnitude higher in floating plants compared to emergent species. In the Tucuruí Reservoir, Aula et al. (1994) also found a much higher Hg content in floating and submerged plants than in emergent species. Plants growing in flooded tailings in Nova Scotia, Canada, also showed the same difference between floating and submerged species and emergent ones (Table 5.1).

Aquatic macrophytes growing on tailings in Canada presented elevated Hg concentrations in shoots, ranging from 0.18 – 0.55 $\mu g\ g^{-1}$ dry wt. (Lane et al. 1988), whereby root concentrations were much higher than in shoots. Lacerda et al. (1991b), studying Hg concentrations in an emergent macrophyte, *Potenderia lanceolata*, along the drainage area of gold mining tailings in central Brazil, found no significant correlation between Hg levels in the plant and in bottom sediments. However, the highest values measured in the plant were generally found associated with high sediment values. They suggested that the plant preferably incorporates Hg^{2+}, and since most Hg exported from those tailings were strongly bound to suspended particles, its availability for plant uptake would vary according to the environmental conditions of the drainage which would control Hg release from suspended particles rather than the bulk Hg concentration in

Table 5.1. Comparison between Hg concentrations in leaves of emergent (e) and floating/submerged (f) aquatic macrophytes from gold mining areas

Study	Hg ($\mu g\ g^{-1}$)	Author
Madeira River SW Amazon, Brazil		Martinelli et al. (1988)
Echinocloa polystachya (e)	0.001	
Victoria amazonica (f)	0.910	
Eichornia crassipes (f)	1.040	
Tucuruí Reservoir, SE Amazon, Brazil		Aula et al. (1994)
Scirpus cubensis (e)	0.030	
Sauvinia auriculata (f)	0.120	
Oldham tailings, Nova Scotia, Canada		Lane et al. (1988)
Sparganium fluctuans (f)	16.30	
Juncus articulatus (e)	0.55	
Juncus pelocarpus (e)	0.54	

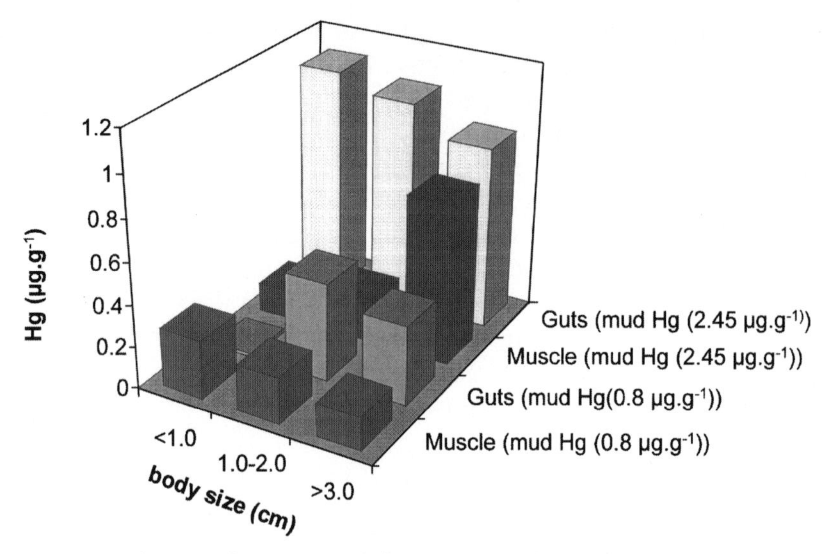

Fig. 5.1. Distribution of mercury in different organs of the freshwater snail, *Hemisinus tuberculatus* Spix of different size, and collected in two differently contaminated sites along the Teles Pires River, southern Amazon. (After Farid 1992)

sediments (Fig. 5.1). In a Siberian gold district, a small amount of discharged metallic Hg was assimilated by aquatic plants found in the local drainage (Taysayev 1991).

Aula et al. (1994) studied the Hg distribution in various submerged and floating macrophytes at the Tucuruí Reservoir, SE Amazon. They found much lower values than those reported for tailings. In all plant species studied, Hg concentrations were significantly higher in roots and submerged leaves than in the shoots and floating leaves. The ratios between Hg concentrations in roots and shoots ranged from 2.67:1 in *Scirpus cubensis*, 2.33:1 in *Eichornia crassipes*, and 1.6:1 in *Salvinia auriculata*.

Plants growing on drainage tailings in Nova Scotia presented root:shoot ratios ranging from 1.04–11.13 (Lane et al. 1988); a ratio of 6.17 was found in the floating macrophyte *Eichornia crassipes* from gold mining sites in the southern Amazon (Farid 1992).

The higher content of Hg in roots relative to shoots in freshwater macrophytes is a well-known phenomenon, which is also found for other trace metals and terrestrial plants (Kelly 1988). Roots act as an efficient barrier to Hg uptake. Mercury can complex to sulfhydryl groups at the root surface (Siegel 1973) and adsorb to deposited iron and manganese plates (Kozlowski 1984; Crowder and StCyr 1991; Gambrell 1994). Also, release of Hg^0 vapor from shoots has been reported for many plant species (Siegel et al. 1980, 1987; Siegel and Siegel 1985), which results in less Hg in the leaves (Table 5.2).

Table 5.2. Comparison between root and shoot mercury concentrations (μg g^{-1}) in aquatic macrophytes from gold mining areas

Species	Shoot	Root	Root: shoot	Author
Equisetum fluviatile, from tailings drainage in Canada	0.45	0.47	1.04	Lane et al. (1988)
Juncus articulatus, from tailings drainage in Canada	0.55	2.69	4.89	Lane et al. (1988)
Juncus pelocarpus, from tailings drainage in Canada	0.54	6.11	11.13	Lane et al. (1988)
Dulichium arundinaceum, from tailings drainage in Canada	0.19	0.67	3.53	Lane et al. (1988)
Leerzia oryzoides, from tailings drainage in Canada	0.18	1.56	8.67	Lane et al. (1988)
Scirpus cubensis, from Tucuruí Reservoir, SE Amazon	0.03	0.08	2.67	Aula et al. (1994)
Eichornia crassipes, from Tucuruí Reservoir, SE Amazon	0.03	0.07	2.33	Aula et al. (1994)
Eichornia azurea, from the Teles Pires River, southern Amazon	0.06	0.37	6.17	Farid (1992)

Seasonality may influence the Hg content in macrophytes. In most tropical regions, the rainy season is associated with high suspended matter (Lacerda et al. 1993). Mercury binds strongly to suspended particles, becoming unavailable for plant uptake. Mercury concentrations in floating macrophytes (*Salvinia auriculata*) during the dry season were twice the average values found during the rainy season (Aula et al. 1994).

5.3 Animals (Excluding Fish)

A few animal species, besides fish, have been analyzed in Hg-contaminated sites from gold mining areas, in order to find biological monitors of Hg contamination. Among them, freshwater snails have been studied more intensively, since these organisms proved to be very good monitors in many situations involving trace metal contamination (MARC 1987a,b).

Freshwater snails are omnivorous animals, feeding on bottom sediments and on plant surfaces. Hg concentrations are therefore generally higher in the guts of these animals than in muscle tissues. Figure 5.1 shows the distribution of Hg in the guts and muscle tissues of *Hemisinus tuberculatus* Spix, a typical and abundant snail from the neotropics, from different sites along the Teles Pires River, southern amazon (Farid 1992). Guts presented much higher Hg concentrations than muscle tissues, and reflect the general Hg content of fine bottom sediments from each site. The mercury content in guts did not correlate significantly with animal size. A positive relationship, however, was found between body size and Hg content in muscle tissue, at the more contaminated site, but not at the less contaminated site.

Hg distribution in *Ampullarius* sp., collected at a contaminated drainage site ofgold mining tailings in central Brazil, ranged from 0.43 $\mu g\ g^{-1}$ in the smallest individuals to 0.95 $\mu g\ g^{-1}$ in the larger individuals. There is a significant, positive correlationship between Hg concentration and shell size ($r = 0.938, p < 0.05$), indicating that size is an important factor determining Hg concentrations in this species. The positive size–Hg concentration relationship has been previously reported for other mollusk species contaminated by Hg (Huckabee et al. 1979; Mohlenberg and Riisgard 1988; Riisgard and Fame 1988), and has been attributed to the accumulation of the long biological residence time of methyl-Hg.

In a study of freshwater snails distributed in the Tucuruí Reservoir, SE Amazon, Aula et al. (1994) found no significant correlations between Hg content and body size or mass. Mercury concentrations were low and quite variable (11 to 170 $\mu g\ kg^{-1}$) and seemed independent of Hg concentrations in bottom sediments.

The results from these two snail species from different mining sites suggest that Hg methylation processes are currently occurring in both areas, at least at sites of elevated Hg content, since Hg concentration in both water and sediments are in general very low, even at the most contaminated drainage sites (Lacerda et al. 1990b; Farid 1992).

Mercury concentrations in relation to sediment content and distance from tailings were determined in the snail *Ampullarius* sp. along a contaminated drainage of the "Tanque dos Padres" tailing deposit. The results are summarized in Fig. 5.2. The results showed that samples collected close to the tailings presented much higher concentrations, decreasing sharply as the distance from the tailings increased, similar to Hg distribution in bottom sediments. A proportional response to environmental Hg levels has been reported for many mollusk and macrophyte species and is one of the main reasons for using these organisms as biological monitors for heavy metal contamination (MARC 1987a, b). The rapid decrease in Hg concentrations in these organisms, however, reflects the low mobility of Hg in the area and also suggests that exported Hg is probably under a low bioavailable form in the drainage.

Aula et al. (1994) studied the distribution of Hg in various animals inhabiting the Tucuruí Reservoir, SE Amazon (Table 5.3). In general all animals presented very low and variable Hg content, in particular the herbivorous

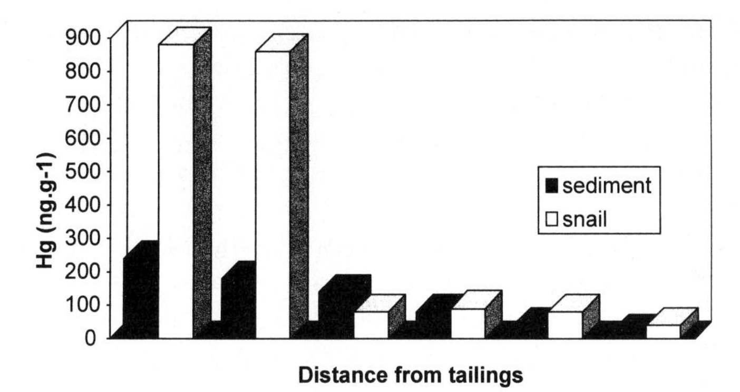

Fig. 5.2. Mercury concentrations in freshwater snails (Ampularius sp.) along a contaminated creek draining tailings deposits in Pocone mining site, central Brazil

Table 5.3. Mercury concentrations (µg kg^{-1}) in the aquatic fauna inhabiting the Turucuí Reservoir. (According to Aula et al. 1994)

Animal	Mean Hg concentration	Range of Hg concentrations
Freshwater snails	57	11–170
Capybara meat	15	7–26
Capybara liver	10	6–12
Turtle eggs	12	7–23
Caiman meat	1900	1200–1600
Caiman liver	19000	11000–30000

species. Mercury concentrations ranged from 15 µg kg^{-1} in snails, to 1900 µg kg^{-1} in caiman meat. Extremely high concentrations (from 11000 to 30000 µg kg^{-1}) were found in the liver of caiman, the main predator in the reservoir.

5.4 Fish

Fish is the main transfer pathway of Hg from a contaminated environment to humans, since this heavy metal typically undergoes biomagnification through food chains, presenting the highest concentrations in high trophic level fish (Moriarty 1974; Huckabee et al. 1979; Johnels et al. 1979). Biomagnification is due to the long residence time of methyl-Hg in animals. Therefore, fish meat is the major source of methyl-Hg to humans. The importance of such pathways within the Hg toxicological cycle has resulted in an extensive amount of data on the Hg content in fish from gold mining areas, since in many of such areas, humans consuming such fish have already presented symptoms of mercury poisioning.

Typical background concentrations of Hg in freshwater fish range from 0.01 $\mu g\ g^{-1}$ wet wt. in short-lived herbivorous species to 0.2 $\mu g\ g^{-1}$ wet wt. in large carnivorous species; although they can reach up to 0.8 $\mu g\ g^{-1}$ wet wt. in large, long-lived marine species such as tuna, swordfish and sharks (Johnels et al. 1979; GESAMP 1986). However, these high Hg contents of large carnivorous oceanic fish may be attributed to anthropogenic Hg emissions to the atmosphere (Fitzgerald et al. 1994; Mason et al. 1994) and to increasing Hg deposition rates over the oceans during the second half of the present century (Slemr and Lander 1992).

In contaminated areas, Hg concentrations in fish reach a few $\mu g\ g^{-1}$ wet wt. or more in carnivorous species. Concentrations higher than 20 $\mu g\ g^{-1}$ wet wt. are considered lethal to fish (Meili 1991a). However, in the Minamata area concentrations up to 300 $\mu g\ g^{-1}$ wet wt. have been reported (Fujiki 1963). Consumption of Hg-contaminated fish is by far the most important transfer pathway of Hg to humans (Mitra 1986; Hacon 1991). This pathway has been proposed as the main route of human Hg intake at many gold mining sites worldwide (Souza et al. 1989). Ingestion of contaminated fishes by humans may be critical when it occurs frequently, because the methylated form presents very high intestinal absorption rates (>95%), high chemical stability, and long residence times in the human organism. Many human populations living in the vicinity of mining areas have contaminated fish as their principal protein source.

The distribution of Hg in fish is generally dependent on the trophic status of a given species, size or age, migratory habits, and the degree of Hg contamination of a given area. In general, higher Hg concentrations are found in the liver of carnivorous fish, since the major Hg uptake pathway is food, followed by muscle tissues where methyl-Hg concentrates. When the uptake involves inorganic Hg species, however, the highest concentrations are found in the gut, kidney and liver, suggesting fast uptake and excretion of Hg (Bargagli 1995).

5.4.1
Mercury Concentrations and Distribution in Fish

Contamination of fish by Hg released in the environment by gold mining activity has been reported in many gold mining areas, and has received more attention from researchers than any other aspect of Hg contamination from this source. Unfortunately, however, most studies were extensive surveys of Hg concentrations in fish, not taking into consideration other important organisms of these ecosystems. A tentative summary of these results is presented in Table 5.4. Not surprisingly, most studies are concentrated in the Brazilian Amazon region, which has over 1300 freshwater fish species, a third of the world's total and the main diet of most local inhabitants (Petrere 1995).

Regardless of site, mining procedures or Hg emissions, carnivorous fish always present the highest Hg concentrations. This distribution is typical of Hg accumulation in fish, either from contaminated or non-contaminat-

Table 5.4. Mercury concentration (μg g^{-1} wet wt.) in fish muscle tissue of fish having different feeding habits. The data, the range of concentrations as they appeared in the original articles, were pooled from different literature sources for a single mining site when existing

Site	Habit[a]	Hg content	Authors
Madeira River, upstream Porto Velho, mining reservation	C NC	0.07 – 2.89 0.02 – 0.65	Martinelli et al. (1988); Malm (1991)
Madeira River, downstream of the mining reservation	C NC	0.67 – 1.47 0.05 – 1.01	Malm et al. (1990); Martinelli et al. (1988)
Poconé region, soil mining and tailings	C NC	0.06 – 0.68 < 0.04 – 0.16	Oliveira et al. (1990); CETEM (1989)
Paraíba do Sul River, SE Brazil	C NC	0.16 – 0.37 0.01 – 0.22	Lima et al. (1990); Pfeiffer et al. (1989a)
San Juan River, Choco, Colombia	C NC	0.66 – 1.26 0.04 – 1.87	CODECHOCO (1991)
Teles Pires River mining region, S Amazon, Brazil	C NC	0.05 – 3.82 0.02 – 0.19	Farid (1991); Akagi et al. (1994)
Tucuruí Reservoir, SE Amazon, Brazil	C NC	0.99 – 2.90 0.05 – 0.60	Aula et al. (1994)
Carajás mining district, SE Amazon, Brazil	C NC	0.11 – 2.30 < 0.01 – 0.31	Lacerda et al. (1994)
Tapajós River, SE Amazon, Brazil	C NC	0.04 – 2.58 0.01 – 0.31	Akagi et al. (1994); Rodrigues et al. (1992)
Lerderderg River, Victoria, Australia	C	0.03 – 0.64	Bycroft et al. (1982)
Davao del Norte, Philippines	C	0.05 – 2.60	Torres (1992)
Negro River, Amazon, Brazil	C NC	? – 4.20 0.14 – 0.35	Malm et al. (1994); Forsberg et al. (1994)
Cuyuní River, Guyana Shield, Venezuela	C NC	0.07 – 0.86 < 0.03 – 0.24	Nico and Taphorn (1994)
Non-contaminated Amazon rivers	C NC	< 0.17 < 0.10	Pfeiffer et al. (1989a)

[a] C, Carnivorous; NC, non-carnivorous from various gold mining areas.

ed sites, and has been extensively discussed in the literature as a result of the biomagnification of alkyl-mercurials, in particular methyl-Hg, which shows fast incorporation and slow excretion rates, resulting in higher Hg concentrations in higher trophic level organisms.

The highest Hg concentrations were found in traditional mining sites located in large Amazonian rivers such as the Madeira River, the larger tributary of the Amazon River. In this river, 25 species of fish were sampled from 1987 to 1991, with a mean Hg concentration of 0.9 $\mu g\ g^{-1}$. Maximum Hg concentrations up to 2.89 $\mu g\ g^{-1}$ in carnivorous fish were reported (Malm 1991). Other traditional mining sites in the Teles Pires and Tapajós River, southern Amazon, showed an average Hg concentration of 0.55 $\mu g\ g^{-1}$, with maximum Hg concentrations reaching 3.82 and 2.58 $\mu g\ g^{-1}$, respectively (Farid 1991; Akagi et al. 1994; Malm et al. 1995). High Hg concentrations were also found in the Itacaiúnas-Parauapebas river system, in the Carajás Mining District in northern Pará State, Amazon, where Hg concentrations in eight fish species ranged from 0.01–2.30 $\mu g\ g^{-1}$ (Lacerda et al. 1994). Mercury concentrations from 48 fish collected along estuaries of Davo Gulf and along the rivers Ngan and Manat in the Philippines, showed Hg concentrations ranging from 0.05–2.60 $\mu g\ g^{-1}$ (Torres 1992). Whereas in the Negro River, central Amazon, even with no mining site operating in this river basin, 50% of 96 fish samples showed Hg concentrations higher than 0.5 $\mu g\ g^{-1}$ (Forsberg et al. 1994; Malm et al. 1995).

These high concentrations reflect the large Hg load to the environment due to gold mining activities. However, they also are a result of the major environmental conditions of most Amazonian aquatic ecosystems which favor high rates of Hg methylation, such as high bacterial activity, slightly acidic conditions of most water bodies and fast turnover and high concentrations of organic matter, and therefore accumulation in high trophic level fish species (Lacerda and Salomons 1984; Lacerda et al. 1989; Forsberg et al. 1994; Guimarães et al. 1994).

In other tropical forest rivers, where gold mining is relatively recent (less than 10 years), Hg concentrations are smaller, such as along the Cuyuní River, Guyana Shield region Venezuela, the San Juan River, and Colombia and the Bolivian rivers along the Brazilian border. Fish from these rivers present maximum Hg concentrations in carnivorous species of 0.86, 1.26, and 0.80 $\mu g\ g^{-1}$, respectively (CODECHOCO 1991; Nico and Taphorn 1994; Zapata 1994). Also, in southeastern Brazilian rivers, with scattered but small-scale mining, Hg concentrations in fish are smaller (0.16–0.37) (Pfeiffer et al. 1989b; Lima et al. 1990), but significantly higher than expected background concentrations. These results strongly suggest that even when relatively small amounts of Hg are released from recent gold mining sites, Hg concentrations in fish are significantly higher than background concentrations.

In areas where major Hg sources are from tailings and soil leaching, Hg concentrations are lower than in tropical forest rivers. In the Poconé region, Mato Grosso State, central Brazil, where intense gold mining occurs mostly in soil, Hg concentrations in carnivorous fish are much lower than those reported for large rivers in the Amazon region, ranging from 0.06 – 0.68 μg g^{-1}.

Mercury inputs to the local environment in these areas are mostly restricted to leaching of tailings, where Hg presents very low mobility (Lacerda et al. 1990b). Therefore, widespread contamination of large water bodies and their fish fauna is not expected. Lower concentrations (0.03 – 0.64 μg g^{-1}) were also found in carnivorous fish from the Lerderderg River, Victoria, Australia, where the major Hg source is also from leaching of tailings (Bycroft et al. 1982). Similarly, in highly contaminated tailings drainage from Hg mining in Monte Amiata, Italy, Hg concentrations in fish were also very low (Bargagli 1995), notwithstanding the presence of Hg drops at the bottom of these drainages.

The lower concentration of Hg in fish from soil mining relative to river mining confirms the higher environmental impact of river mining. Souza et al. (1989) estimated risk factors of human contamination through fish consumption to be three times higher in river mining sites in the Amazon region than in two other sites in Rio de Janeiro and Para State. However, the specific characteristics of human diet and Hg availability in fish hamper any generalizations regarding the situation where Hg is most available for human intake.

Extreme Hg concentrations in large carnivorous fish from gold mining areas in the Amazon are compared with data from fish from other Hg-contaminated sites by industrial effluents in the world (Table 5.5).

Table 5.5. Extreme Hg concentrations in large carnivorous species from gold mining sites in the Amazon and from different contaminated areas in the world

Species	Site	Hg μg g^{-1} net weight	Author
Paulicea lutkeni Steindachner	Carajás mining site	2.19	Fernandes et al. (1989)
Pseudoplatistoma fasciatus L.	Madeira River	2.70	Pfeiffer et al. (1989)
Brachyplatystoma filamentosum Lichtenstein	Teles Pires River	3.82	Akagi et al. (1994)
Esox lucius L.	Canadian lakes	2.87	Olgivie (1991)
Mullus barbatus L.	Tyrrhenian Sea, Italy	2.20	Bacci et al. (1990)
Esox lucius L.	Finish lakes	1.80	Mannio et al. (1986)

The results show that Hg concentrations in fish from gold mining areas are in the same order of magnitude as those found in areas contaminated with Hg from industrial uses, such as in the Great Lakes, and also in fish from areas under the influence of Cinnabar deposits such as the Tyrrhenian Sea, Italy.

5.4.2 Methyl-Mercury in Fish from Gold Mining Areas

It is clear from the Hg concentrations in fish from surveyed areas that carnivorous fish presented the highest concentrations of Hg in any region studied, frequently surpassing the maximum permissible concentration of 0.5 $\mu g\ g^{-1}$ for Hg in fish for human consumption (WHO 1976). This pattern of Hg distribution is typical of the methylated Hg complex (Fowler et al. 1978; Mitra 1986).

Indirect evidence of Hg methylation in gold mining environments is also shown by the relationship of Hg concentration and fish size. Methyl-Hg shows very slow excretion rates, tending to accumulate in large, old individuals (Moriarty 1974; Mohlenberg and Riisgard 1988; Riisgard and Famme 1988).

Lacerda et al. (1994) showed Hg distribution in two top predators: the jau (*Paulicea luetkeni*) and the piranha (*Serrasaumus nattereri*) in the Carajás mining region. For both fish species, Hg contents were significantly higher in larger, older individuals, generally following a logarithmic pattern. These curves strongly suggest that methylation is occurring in the river systems.

Methyl-Hg was directly measured in 26 fishes from the Tapajós River, southeastern Amazon, and showed an average percentage of around 90% (range 65–100%) of the total Hg content. This value can be considered low if compared with preliminary data from the Madeira River, where methyl-Hg reached 98% of the total Hg concentration (Akagi et al. 1994).

Padberg (1990), in her study of the same river, found low methyl-Hg concentrations (38–563 $\mu g\ kg^{-1}$), corresponding to 60 to 100% of the total Hg content of the studied fish. It is important to note that in the same area, the methyl-Hg content in water reaches only 10% of the total concentration and only 2% of the total Hg concentration found in sediments. This reflects the higher biological residence times of methyl-mercury in fish, and that even under very low Hg environmental concentrations, this phenomenon can easily result in high Hg levels in carnivorous fish, , in agreement with most literature data, which also confirm that most Hg present in fish muscle is in the form of methyl-Hg. Unfortunately, methyl-Hg data are not available for other major gold mining sites outside the Amazon.

5.4.3
Fish Response Changes in Hg Environmental Concentrations

5.4.3.1
Spatial Variability

As has been shown for other aquatic organisms, fish also respond to spatial variability of Hg concentrations, at least to a limited scale. Spatial variability has been only reported for sites where a point source of Hg exists, such as tailings. In large tropical rivers, a decreasing Hg concentration in fish with increasing distance from mining operations is not observed. Fish from the Madeira River, collected at Humaitá and Manicoré (mining areas 180 and 500 km downstream, respectively), presented concentrations similar to those from fish sampled closer to mining operations (Malm 1991). This suggests that mercury became widely and rather uniformly dispersed in the environment.

Bycroft et al. (1982) studied the concentration of Hg in two benthic feeder fish species along the Lerderderg River, which had accumulated Hg from last century tailings. The two species, the brown trout (*Salmo trutta*) and the blackfish (*Gadopsis marmoratus*), responded differently to the Hg concentrations in sediments. Mercury concentrations in the brown trout, as a vagile species that migrate downriver when reaching adulthood, did not correlate well with sediment Hg levels. Blackfish, however, a more sedentary species, showed the highest level where Hg in sediment was also highest. Such sedentary species can be expected to be better biological indicators of Hg-contaminated sediments. Larger fish of both species presented significantly higher Hg concentrations than smaller fish (Fig. 5.3).

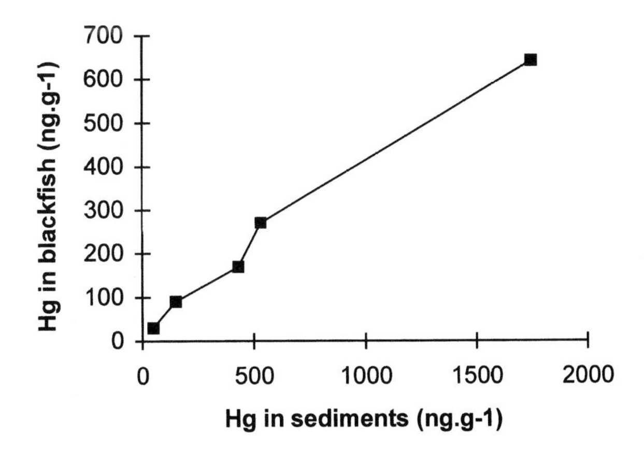

Fig. 5.3. Relationship between Hg concentrations in blackfish (*Gadopsis marmoratus*) and in bottom sediments from the Lederderg River, Australia. (After Bycroft et al. 1982)

5.4.3.2
Temporal Variability

Some studies related Hg content in fish with season. In the tropics, although temperature is not a significant seasonal variable, rainfall and consequent hydrological variability may change Hg availability for fish uptake. For example, Zapata (1994) reported seasonal variation in Hg content of fish from Bolivian rivers. Mercury concentrations average 0.58 $\mu g\ g^{-1}$ in the rainy season and 0.80 $\mu g\ g^{-1}$ in the dry season. More intense gold mining during low river level and higher remobilization of deposited Hg in bottom sediments were suggested as major causes for the seasonality found. Along the Itacaiúnas-Parauapebas rivers in the Carajás region, total Hg content in waters increased from 111 $ng\ l^{-1}$ in the rainy season to 320 $ng\ l^{-1}$ river volume, resulting in larger input and less dilution of Hg (Fernandes et al. 1991).

A detailed study on Hg variation through time in carnivorous fish species was carried out in this river system (Fernandes et al. 1991). The authors studied over 3 years the Hg concentration of local fish fauna dominant in the region and important as food items of the local population. Table 5.6 shows a summary of their results. Highest concentrations of Hg were found in carnivorous fish as expected, and among them the larger (> 20 kg) species showed extremely high concentrations, always higher than 1.0 $\mu g\ g^{-1}$ wet wt., and as high as 2.3 $\mu g\ g^{-1}$ wet wt. in the large carnivorous *Paulicea lutkeni*. A comparison of the results obtained in 1988 and 1990 strongly suggests that Hg concentrations are increasing, in particular in carnivorous species, simultaneously with increasing gold production in the area.

Table 5.6. Mercury concentrations ($\mu g\ g^{-1}$ wet wt.) in fish from the Carajás mining site collected between 1988 and 1990

Fish species[a]	Hg ($\mu g\ g^{-1}$ wet wt.)	
	1988	1990
Paulicea lutkeni (C)	0.80–1.46	1.25–2.30
Prochilodus nigricans (H)	0.13–0.31	0.01–0.02
Pimelodus sp. (C)	0.09–0.24	0.17–0.19
Brycon sp. (H)	0.05–0.16	0.04
Serrasalmus nattereri (C)	0.10	0.01–0.87
Leporinus sp. (H)	0.01	0.01–0.03
Hoplias malabaricus (C)	0.35–0.91	0.31

[a] C, Carnivorous fish; H, herbivorous fish.

The available data on Hg concentrations in fish from gold mining sites worldwide are sufficient to confirm a generalized contamination of fluvial environments. In various sites, Hg concentrations are nearly five times the maximum permissible concentrations for human consumption. Other important organisms for aquatic food chains like freshwater snails and aquatic macrophytes also presented elevated Hg concentrations, indicating widespread contamination of these areas. The complex nature of the interactions between components of tropical food chains makes it imperative to gather data on Hg concentrations in other important carnivores, such as aquatic mammals and reptiles, and of birds, in order to have a more complete framework of Hg transfer through food chains.

5.5 The Atmospheric Link in the Mercury Content of Fish

Fish from large floodplain rivers, black water rivers, and artificial reservoirs in the tropics show the highest Hg content when compared to other freshwater systems. This occurs regardless of the magnitude of the direct Hg inputs to a given river or reservoir, or of the large dilution of Hg in these large-volume water systems. For example, the Hg content in piranhas (*Serrasalmus* sp.) from the Tucuruí reservoir, SE Amazon, which has no direct input of Hg, is among the highest reported for this species in the Amazon region, showing concentrations typically higher than 3.0 $\mu g\ g^{-1}$ (Aula et al. 1994). Fish from remote areas of the Negro River basin, central Amazon, also show a very high Hg content, with values as high as 2.0 and 3.0 $\mu g\ g^{-1}$ (Forsberg et al. 1994).

Mercury concentrations in fish do not correlates with the total Hg present in water. However, significant correlations were found between dissolved organo-Hg and total Hg content in fish meat collected from many river and lake systems in California with different Hg concentrations and speciation in the water (Gill and Bruland 1990; Fig. 5.4). Thus the methylation capacity of a given river or lake environment will actually control Hg concentrations in fish, rather than the total Hg load or concentration in water.

For example, fish from Hg mining areas, where drops of metallic Hg are frequently found on the bottom of creek sediments, but with waters very poor in methyl-Hg, present an extremely low Hg content (Bargagli 1995).

Floodplains, black water rivers, and tropical reservoirs present various biogeochemical characteristics which favor Hg methylation. These freshwater systems are rich in oxidizable organic matter and nutrients, which facilitate microbial processes. Furthermore, the acidity and low conductivity of these waters (typically $pH < 6.0$; conductivity $< 50\ \mu S\ cm^{-1}$) create

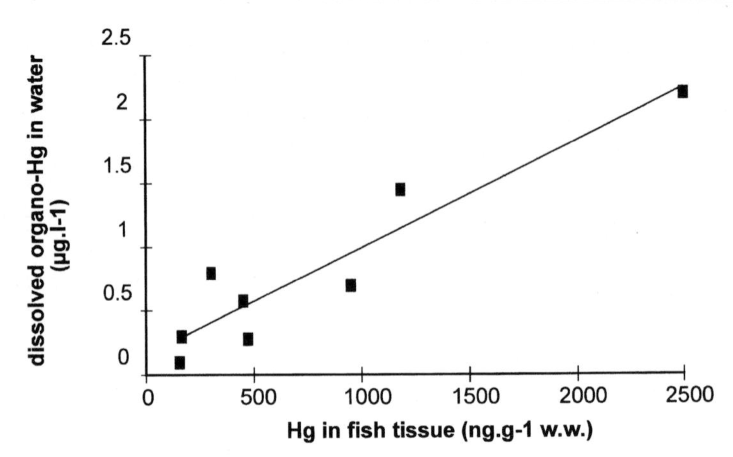

Fig. 5.4. Dissolved organo-Hg (ng l^{-1}) concentrations and total Hg concentrations in fish meat from various freshwater systems in California. (After Gill and Bruland 1990)

ideal conditions for Hg methylation (Bloom et al. 1991). Also, high concentrations of dissolved organic matter enhance the solubility and stability of Hg in water through complexation.

The coupling of Hg and high concentrations of organic matter has been suggested as a major cause of fast production of methyl-Hg in many aquatic systems such as new, ponded reservoirs (Jackson 1988); forest lakes (Meili 1991a–d; Meili et al. 1991); floodplains (Lacerda et al. 1989), and black water rivers (Forsber et al. 1994). Elevated Hg contents in fish from remote seepage reservoirs have been extensively reported, notwithstanding the absence of any significant Hg source.

Mercury methylation rates measured in various Amazon sites, for example, were always higher in floodplains, black water rivers, and reservoirs by a factor of 100 to 10 000, when compared with white water rivers and other neutral pH, low organic matter and high conductivity freshwater systems (Guimarães 1992; Guimarães et al. 1993a, b, 1994a, b).

These data suggest that a high Hg content in fish from large floodplain rivers and reservoirs can be partially explained by the high rates of methylation. However, where is the source of Hg for methylation? In the Madeira River, a white water river with extensive floodplains, for example, direct Hg inputs may reach up to 40 tons $year^{-1}$. At the Negro River, there are virtually no direct Hg inputs. Fish from both rivers, however, present high and similar Hg concentrations. Fish from white and clear water rivers in central Brazil, with small floodplains, although receiving direct Hg inputs from mining operations, show a very low Hg content. Direct inputs from mining consist of metallic Hg (Hg^{0}), which does not serve as a substrate for methylation. Methylation requires a donation of a methyl radical and Hg^0 is unlikely to be methylated via a cobalamin pathway, as is Hg^{2+} by sulfate bacteria for example (Mason and Morel 1993). Mercuric ions (Hg^{2+}) are the fundamental source of Hg for the bacterial methylation process.

Sources of Hg^{2+} could be the oxidation of Hg^0 in the water column or the deposition of reactive Hg formed by atmospheric oxidation of Hg vapor. Also, under the conditions of amalgam burning or gold purification, Hg^0 can be emitted. Since Hg^0 is practically nonreactive in oxic natural waters, the major source of Hg^{2+} to most tropical environments is the atmosphere, where Hg vapor is rapidly oxidized and washed out under the conditions prevailing in such atmospheres (see Chap. 3 for details).

Results from Hg distribution in lake sediment profiles (Lacerda et al. 1991), from Hg distribution between vapor and particulate phases in the atmosphere (Hacon et al. 1995), and fromthe speciation of Hg emitted to the atmosphere during roasting of the amalgam (Marins and Tonietto 1995) suggest that the residence time of Hg in the atmosphere of gold mining areas should be days. Deposition rates measured in gold mining areas also suggest fast atmospheric oxidation and washout and may reach up to 100 $\mu g\ m^{-2}\ year^{-1}$. Therefore, most of the Hg emitted to the atmosphere returns to soil and aquatic environments as Hg^{2+}.

6 Mercury Contamination of Humans in Gold and Silver Mining Areas

Human contamination by Hg in gold and silver mining has been reported from various sites throughout the world, since the large-scale use of Hg was introduced as the major mining technique to produce silver in the Spanish colonial Americas (Galeano 1981). Present-day surveys carried out in many countries have shown that contamination is mostly reflected by higher Hg concentrations in body indicators (e.g., hair, urine, and blood). Nearly all surveys carried out at gold and silver mining sites worldwide have shown these concentrations to be higher than expected background levels. However, only a few studies had actually detected symptoms or clinical signs of Hg poisoning. Most health authorities agree that the lack of information on "mercurialism" among the exposed population is probably due to difficult logistics and the poor health conditions of the population which may mask symptoms of mercury poisoning.

Difficult logistics and lack of reliable medical information regarding most remote populations have hampered a better understanding of Hg contamination in humans from a truly epidemiological point of view. Therefore, in this present review, we will restrict ourselves to the presentation and discussion of the existing data on Hg concentrations in humans.

In May 1987, the first Hg poisoning incident in Tagun, Davao del Norte, the Philippines, was reported. Eleven persons were reported ill and one died in this incident, when approximately 2 kg of gold amalgam were processed (Torres 1992). In the Brazilian Amazon, cases of mercurialism have been reported, although no deaths could be definitively linked to Hg poisoning (Branches et al. 1993). Notwithstanding the few existing and confirmed data, it is our strong belief that a silent outbreak of Hg poisoning is turning rapidly into a regional disaster.

6.1
Exposure Characteristics to Human Groups

Human exposure to Hg from gold and silver mining is basically through two main routes. The first main pathway involves occupational exposure. This affects miners who burn the amalgam in the field and the people working at gold dealer shops. This latter group includes those directly involved in the roasting of the ore for purification, and those involved with other administrative tasks of gold commercialization and who spend at least 8 h/day in these shops (indirect exposure). In China and in certain remote sites in the Amazon, workplaces are also used commonly as living quarters for miners, who are therefore permanently exposed to a Hg-rich atmosphere.

The second main pathway involves environmental exposure through the inhalation of Hg-contaminated air and affects urban populations living close to gold dealer shops and rural populations through the ingestion of contaminated fish. This second group also includes various Indian and peasant populations not directly involved in gold mining and not receiving any profit from gold exploration.

It is well recognized that humans exposed to Hg through the occupational pathway are basically exposed to inorganic Hg, mostly elemental Hg^o, from amalgam burning and roasting of gold for purification. On the other hand, environmental exposure consists mostly of ingestion of contaminated fish, the major form of which is methyl-Hg (Hacon 1991; Akagi et al. 1994).

The difference in Hg exposure from these two pathways will result in different intoxication patterns, symptoms, and monitoring programs and subjects. Blood and urine have been used to monitor environmental and occupational exposure, respectively, and have been shown to be very reliable for application in short-term studies. Cleary (1994) studied the relationship between Hg concentrations in urine and blood and amount of gold burnt (occupational exposure) and daily meals, including contaminated fish (environmental exposure), in a population along the Tapajós river basin, a traditional mining site in the Brazilian Amazon. Blood Hg levels were significantly correlated with the number of days involving a fish meal per week ingested by the population. However, blood Hg concentrations did not correlate with the amount of gold burnt per week. Urine levels showed a different response, being significantly correlated with the amount of gold burnt per week, but not with the number of meals containing fish. The examples clearly confirm urine Hg concentrations as the best monitor of occupational exposure, while blood Hg concentrations best monitor environmental exposure, at least for short- to medium-term

Table 6.1. Average mercury concentrations in urine and blood samples from different groups living along the Tapajós River valley. (n = number of subjects analyzed) (After Cleary et al. 1994)

Group	Hg in blood ($\mu g\ l^{-1}$)	Hg in urine ($\mu g\ l^{-1}$)
Gold traders	30.4	78.9
Miners	34.3	18.5
Resident (adults)	20.0	18.5
(children)	14.5	18.6
Fishermen (adults)	96.7	13.9
(children)	70.4	0.4

surveys. For long-term surveys and for the study of the progress of Hg contamination through time, human hair has been proposed as the best choice for monitoring purposes (Hacon 1991; Akagi et al. 1994), and will be discussed in detail further in this chapter.

In another study in the same area, Cleary et al. (1994) analyzed blood and urine samples from 99 individuals belonging to 4 different groups: traders, miners, fishermen, and resident non-fishermen (Table 6.1). The four groups presented different Hg levels. Gold traders, who are occupationally exposed to Hg vapor during gold distilling, showed the highest Hg concentrations in urine, whereas the highest Hg concentrations in blood samples were found among fishermen, suggesting that blood levels reflect intake through food rather than inhalation of Hg vapor. Prospectors and non-fishermen along the river valley presented relatively low and similar Hg concentrations in blood and urine samples. The results justify detailing the two contamination routes separately.

6.2 Mercury Concentrations in Humans from Occupational Exposure

Gold dealer shops, where the bullion is purified at high temperatures, are without doubt the major point source of atmospheric mercury in most mining areas and are probably responsible for most human contamination in these regions (Hacon 1991). Table 6.2 summarizes Hg concentrations in the ambient air of several shops located in gold mining areas.

The samples collected inside gold dealer shops and workplaces (Table 6.2) showed very high and variable Hg concentrations as expected, ranging from 5.5–106.5 $\mu g\ m^{-3}$ in Poconé, central Brazil (Marins and Tonietto 1995) to up to 292 $\mu g\ m^{-3}$, in Porto Velho, NW Amazon (Malm et

Table 6.2. Mercury concentrations in ambient air of gold dealer shops and workplaces in gold mining areas (in μg Hg m^{-3})

Site	Hg (μg m^{-3})	Author
Davao del Norte, Philippines	42 – 1664	Torres (1992)
Dixing, China	3.3 – 9.9	Yshuan (1994)
Dixing, China	1000 – 2600	Yshuan (1994)
Indonesia	85.0	Achmadi (1994)
Poconé, central Brazil	5.5 – 39.8	Marins et al. (1991)
Poconé, central Brazil	6.8 – 106.5	Marins and Tonietto (1995)
Porto Velho, W. Amazon	Up to 292.0	Malm et al. (1990)
Porto Velho, W. Amazon	< 1.0 – 107.2	Malm et al. (1991)
Alta Floresta, S. Amazon	0.2 – 40.6	Hacon et al. (1995)

al. 1990) and to extremely high concentrations of up to 2600 μg m^{-3} at workplaces involved in amalgam burning in Dixing Province, China (Yshuan 1994). In Davao del Norte, the Philippines, Torres (1992) found Hg concentrations in workplaces ranging from 42 – 1664 μg m^{-3}. Also, among 567 samples collected in four different workplaces, 65 % of them presented concentration higher than the limit of 50 μg m^{-3} set by WHO (1976). The concentrations probably reflect the shop size and architecture, amount of gold commercialized daily, and ventilation. All the reported values were higher than maximum permissible concentrations for public exposure of 1 μg m^{-3} and frequently even higher than the limits for industrial exposure of 50 μg m^{-3} (WHO 1976).

A detailed study on the fate of mercury emitted by a gold dealer shop was done in Pocone, central Brazil (Marins et al. 1991). A collection of 29 samples progressively further from a point source was done taking into consideration wind direction and city architecture. The results showed that the initial total concentration of Hg in air at the door of the shop ranged from 1.6 – 2.3 μg m^{-3}. Samples collected within a100 m radius of the shop showed Hg concentrations ranging from 0.14 – 1.68 μg m^{-3}, which is in the same order of magnitude of values found over cinnabar mines (Friberg and Vostal 1972) and similar to Hg concentrations found by Pfeiffer et al. (1991) in Porto Velho, Rondônia, another gold-producing urban area in the Amazon.

Samples collected within a 450-m radius of the shop showed much lower mercury concentrations, ranging from < 0.14 to 0.17 μg m^{-3}, indicating that most of the Hg emitted was deposited very near the source.

These results seem to describe a typical behavior of Hg emitted to the atmosphere by anthropogenic point sources (Ferrara et al. 1982; Lindqvist and Rhode 1984). Although the small fraction dispersed to larger distances can be sufficient to overpass background deposition rates, even at distances up to 100 km from the source, as has been demonstrated by the relatively elevated Hg concentrations measured in surface sediments of remote lakes relatively far from mining sites (Lacerda et al. 1991b).

Mercury concentrations in the ambient air of workplaces have been studied in many industries and show a wide range of values. Thus workers have been exposed to varying degree of contamination at their places of employment.

In one of the very first studies on Hg contamination of the atmosphere of workplaces, Bidstrup et al. (1951) reported Hg concentrations ranging from 5–200 μg m^{-3} in a workshop that repaired electric current meters. They also reported marked seasonality in Hg concentrations. In another early study on the occupational exposure to atmospheric Hg, Benning (1958) reported Hg concentrations in ambient air ranging from 200–7500 μg m^{-3} in a workroom atmosphere of a company using copper amalgam in manufacturing carbon brushes for electromotors. The room lacked any sanitation or engineering hygiene practice. Introduction of ventilation control measures reduced Hg concentrations to 50–70 μg m^{-3}.

In a jewelry manufacturing company, peak Hg concentrations in summer months, prior to the installation of proper ventilation systems, reached 350 μg m^{-3}, and dropped to nearly 30 μg m^{-3} after the systems were implemented (Copplestone and McArthur 1967). Dental surgery workplaces, on the other hand, had Hg concentrations ranging from 1.5–3.6 μg m^{-3} (Nilsson et al. 1990). Mercury concentrations in the ambient air of chlor-alkali plants in the USA and Canada in the 1960s typically ranged from 1.0–2640 μg m^{-3} (Smith et al. 1970). The ambient air of the last Hg-cell chlor-alkali plant in Brazil had Hg concentrations rangeing from 1.0–64 μg m^{-3} (Calazans et al. 1993).

The concentrations of Hg in gold mining areas or gold commercialization shops (Table 6.2) are in the same range of concentrations found in heavily contaminated industrial workplaces. Therefore these places should be viewed as an industrial site and submitted to the same legislation which rules and controls Hg levels in workplaces, aiming to achieve an acceptable Hg level to which workers can be exposed.

Emissions in indoor areas in cities or villages with poor air circulation are critical for Hg^0 human contamination, occurring mainly through inhalation. Surrounding environments can also be contaminated. However, in these situations it is possible to introduce procedures for Hg recovery by using retorts in gold mining areas and efficient exhaustion systems to collect Hg^0 vapor in gold dealer shops. Despite the small amount of Hg reburned in the shops (around 5% of total losses), when compared to the total Hg used in the whole mining process, these shops represent a potential risk of contamination for people occupationally exposed, as well as for those living in the vicinity. Atmospheric Hg inside these workplaces typically reaches 100 μg m^{-3} and frequently reaches up to 300 μg m^{-3} and higher (see Chap. 4). Even outside the burning places, total Hg concentrations in air can reach from 10–200 μg m^{-3} (Malm et al. 1990).

Inhaled Hg^{o} shows high absorption through the lungs (more than 85%) and, after some time in the bloodstream, it is partly oxidized and accumulated in the kidneys. Mercuric ion is excreted through urine which is the best indicator of Hg exposure and inorganic Hg levels. Typically, the methyl-Hg content in urine corresponds to less than 0.5% of the total Hg concentration (Akagi et al. 1995). Reburning of amalgams (bullion) in gold dealer shops, in cities, and villages contaminates indoor areas and the vicinities, thus representing the critical occupational exposure pathway for Hg^{o}. Levels associated with development of mercurialism symptoms are ca. 100 mg l^{-1} (WHO 1980).

A summary of the most consistent data on Hg concentrations in the urine of exposed human groups reported for many gold mining areas worldwide is summarized in Table 6.3.

Table 6.3. Mercury concentrations in human urine samples from different mining sites and groups occupationally exposed to mercury emissions from gold mining

Study	Hg concentrations (μg l^{-1})	Subject
Barbosa et al. (1995)	average: 14, n = 194	Kayapó indians
Barbosa et al. (1995)	average: 25, n = 109	Prospectors in Rondônia
Branches et al. (1992)	10–1168 urine (average: 250)	Employees of gold traders shops. Itaituba, SE Amazon
CODECHOCO (1991)	<0.01–21 urine	Inhabitants of Adagoya, Choco, Colombia
Achmadi (1994)	3.5–11.3 urine	Workers in gold-room from mines in Indonesia
Câmara (1994)	14.7–160.0 urine	Workers in gold traders shops in Alta Floresta, S Amazon
Branches et al. (1993)	61.0 urine (n = 5)	Workers in gold traders shops with over 5 years of exposure
Cleary et al. (1994)	78.9 urine (n = 42)	Gold traders at the Crepori mines, SE Amazon
WHO (1980)	100	Minimum concentration before developing mercurialism symptoms
WHO (1991)	50	Maximum acceptable concentrations

Urine samples have been used to investigate prospectors, particularly those exposed to Hg^{o} in the shops and workplaces. In a comprehensive study of the Tapajós River mining area, SE Amazon, 64 individuals were investigated, nearly two thirds of the employees from gold dealer shops and the remaining from Santarem city, a large Amazonian town not directly involved with gold mining (Malm et al. 1995).

Highest average values were observed in persons working indoors with little air ventilation or in reburning rooms with air-conditioning. Mercury concentrations in urine from the people exposed in shops were usually high, ranging from 8.5–1168 $\mu g\ l^{-1}$, with an average value of 228 $\mu g\ l^{-1}$ in a total of 75 urine samples analyzed. Workers or people exposed during burning in open areas at mining sites, despite the much larger amounts of Hg manipulated here than in the shops, had much lower urine Hg values, ranging from 0.3–74.3 $\mu g\ l^{-1}$, with an average of 12.4 $\mu g\ l^{-1}$ over 37 samples analyzed (Malm et al. 1995). The maximum acceptable concentration in urine is 50 $\mu g\ l^{-1}$ (WHO 1991), whereas the minimum concentration believed to cause symptoms of mercurialism is 100 $\mu g\ l^{-1}$ (WHO 1980). The concentration in the urine of people exposed in gold dealer shops (Table 6.3) is quite high and this group is obviously the critical group regarding Hg^{o} inhalation.

In Indonesia, a survey of gold mining workers showed the lowest urine Hg concentrations in drivers that served the facility (0.2 $\mu g\ l^{-1}$), whereas urine levels in workers exposed to blowtorches in the "gold room" where amalgam is burnt and gold distilled showed maximum values (25 $\mu g\ l^{-1}$) (Achmadi 1994).

Branches et al. (1992) surveyed workers in gold dealer shops in various areas of the Tapajós mining site, Pará State, eastern Amazon. Out of 45 individuals, they found mercury concentrations in urine ranging from 10–1168 $\mu g\ l^{-1}$, with an average of 250 $\mu g\ l^{-1}$. Mercury concentrations in the ambient air of these shops varied from 7.18–107.8 $\mu g\ m^{-3}$. However, no correlation has been found between mercury concentrations in humans and the ambient air of the shops.

Individual responses such as sensibility or susceptibility were sometimes considered more important factors than the personnel employed in the shop, that is, administrative workers sometimes present similar or higher Hg values in urine as occupationally exposed people. Typical symptoms observed were dizziness, headache, palpitations, tremor, pruritus and insomnia (Branches et al. 1993). In a similar study in the Philippines, another symptom significantly associated with exposure to Hg vapor in workplaces was retardation of gross motoricity (Torres 1994). In the study, 230 workers exposed to inorganic Hg were submitted to physical examination. Of those, 196 (47%) of the subjects showed gray focal deposits in the gingiva. However, only 13 (6%) showed elevated blood Hg levels and urine Hg levels were always below 50 µg l^{-1} (Torres 1994).

In another study along the Tapajós River, one of the better human population studies on Hg contamination, people having different occupations were investigated (Cleary 1994). The results showed that highest Hg concentrations in urine samples came from people involved with burning amalgam. Maximum Hg concentrations among this group reach over 800 µg l^{-1}. Individuals belonging to riverine populations not involved in gold mining as well as the miners themselves presented low Hg concentrations.

Also along the Tapajós River, Barbosa et al. (1995) compared Hg in the urine of prospectors to Hg levels found in Kayapó Indians living in the same area. Among 109 prospectors, 38% showed urine Hg higher than 20 µg l^{-1}, whereas 30% of the Indian samples showed this level. Furthermore, average Hg concentrations in the Indian population were significantly lower (14 µg l^{-1}) than in the group of prospectors (25 µg l^{-1}).

All available results show that Hg in urine indicates a high degree of occupational exposure to Hg. The use of equipment to avoid Hg emissions originating from the burning of amalgam or gold purification is quite restricted or nonexistent. Lack of wareness of risks, disregard, and even the unavailability of equipment are the main difficulties. Efficiency and durability of such devices are still an issue in development. Measurements of Hg concentrations in air during the operation of such equipment showed not only failures in the retorts, but also in the exhaustion hood systems. As expected, most cases of Hg poisoning reported from gold mining areas resulted from occupational exposure involving inhalation of Hg vapor. Although particularly serious, it seems to be very restricted to a specific critical group including amalgam or bullion burners and administrative workers at gold dealer shops. Environmental exposure, on the other hand, seems to affect a much larger population, including those not necessarily involved in gold mining.

6.3 Mercury Concentrations in Humans Due to Environmental Exposure

Environmental exposure to Hg is basically due to the ingestion of contaminated food, in particular of fish. Human levels of exposure are generally assessed through the analysis of blood and hair samples. By using blood samples, the exposure to an acute dose of Hg can be determined. The use of hair samples, on the other hand, gives an estimate of the accumulation of longer periods (months to years).

Environmental exposure has been studied in various mining areas, unfortunately, however, only in the Brazilian Amazon. Due to the large atmospheric inputs of Hg from gold mining, Hg contamination is rather regional, and contamination is spread to areas even far from mining sites. This results in generalized contamination of natural resources, in particular of fish, thus possibly affecting people not directly involved with gold mining. A typical riverine population of that region may ingest up to 500 g of fish per day, which may result in increased Hg levels (Ayres 1994; Barbosa et al. 1995).

6.3.1 Mercury Concentrations in Human Blood

Mercury concentrations in human blood is considered to be correlated with recent intakes of Hg through contaminated food. However, if only inorganic Hg in blood is considered, blood Hg levels can also be related to occupational exposure through inhalation. At least for humans working on a daily basis at gold dealer shops, a positive correlation between total inorganic Hg in blood and in urine was found (Akagi et al. 1995).

Estimates of blood Hg levels considered toxic are quite variable depending on the literature. However, most studies agree that blood Hg concentrations are almost always below 50.0 $\mu g\ l^{-1}$ in unexposed individuals. Symptoms typically begin to develop with blood Hg levels of 20.0–50.0 $\mu g\ l^{-1}$. Although whole blood Hg levels are considered the best measure of acute Hg absorption, the correlation with toxicity in chronically exposed individuals is variable (Florentine and Sanfilippo 1991; Branches et al. 1993).

Cleary (1994) compared Hg concentrations in different groups along the Tapajós River valley regarding their Hg concentrations in blood. The highest blood levels were found in residents, living along the river and with fish as an important food item. Miners and burners of amalgam, although permanently exposed to Hg vapor, presented much lower concentrations in their blood. The data suggest a strong correlation between Hg in blood and fish consumption.

Methyl-Hg is the major form of Hg in blood samples. Significant positive correlations have been found between total and methyl-Hg contents in blood samples (Akagi et al. 1995). Analysis of Hg speciation in blood samples from four mining towns in SE Amazon, Brazil, showed methyl-Hg concentrations ranging from 90–149 $\mu g\ l^{-1}$ ($n = 59$). These concentrations corresponded to 72 to 99% of the total Hg content in blood samples (Akagi et al. 1995).

6.3.2 Mercury Concentrations in Human Hair

Human hair is widely accepted as the best indicator for the assessment of contamination in populations exposed to methyl-Hg. Mercury concentration in hair is significantly correlated with blood concentration when hair is being formed, and may accumulate in the hair for periods of months to years. The hair to blood ratio of mercury is in general 250:1. This makes this indicator ideal for monitoring purposes (Akagi et al. 1995). In general, methyl-Hg corresponds to 80 to 100% of the total Hg content in hair samples.

The Hg content of hair is also related to the daily intake of Hg and with the first symptoms of Hg intoxication. Therefore, this indicator may be successfully used in monitoring and risk assessment programs. Table 6.4 shows some indicators of health effects, daily Hg intake, and the respective Hg concentrations in human hair, which can easily be assessed in gold mining areas.

Taking into consideration the Hg concentrations reported for gold mining areas (see Table 6.5), one can conclude that many groups in these areas already show Hg concentrations in hair at which health effects can be expected.

In a survey of Hg concentrations in hair of the population working in mining sites along the Madeira River, Malm (1992) showed a median value of 4.6 μg g^{-1} with concentrations ranging from 0.2–24.1 μg g^{-1}. On the other hand, among the population not directly involved in gold mining, this author found much higher values, ranging from 0.5–71.3 μg g^{-1} with a median value of 7.9 μg g^{-1} (Table 6.5).

Malm et al. (1992) analyzed Hg concentrations in 56 persons, having a fish diet, along the Tapajós River mining area, but who were not directly involved in gold mining. Hair analyses from this population were compared to samples from an urban population of Santarem city, eastern Amazon, which also was not directly involved with gold mining. Fish eaten by the two groups were also analyzed. Mercury concentrations in fish consumed by the riverine population were significantly higher than those measured

Table 6.4. Methyl-Hg concentrations in hair and related daily Hg intake and expected health effects in humans, assuming an average body weight of 55 kg. (After Barbosa et al. 1995; Boichio et al. 1995)

Indicator	Methyl-Hg in hair (μg g^{-1})	Daily Hg intake (mg kg^{-1})
FAO/WHO maximum allowable concentration	7.0	0.5
EPA reference dose	4.0	0.3
Abnormal infant development	10–20	0.7–1.5
Paraesthesia	50–100	160–380

Table 6.5. Mercury concentrations in human hair samples from different mining sites

Author	Hg concentration ($\mu g\ g^{-1}$)	Observations
Hentschel et al. (1992)	< 0.01 – 17.1	Prospectors in the Nariño Province, Colombia
Malm (1992)	0.2–24.1 (4.6)	Prospectors along the Madeira River, Rondônia, W Amazon
Malm (1992)	0.5 – 71.3 (7.9)	Riverine population having fish as the main diet along Madeira River
Malm et al. (1992)	0.74 – 17.7	Riverine population having fish as the main diet along Tapajós River
Forsberg et al. (1994)	5.76 – 171.4 (75.5)	Riverine population along the Negro river, central Amazon, where no gold mining occurs
Aula et al. (1994)	4.0 – 241 (47)	Riverine population around the Tucuruí reservoir, where no gold mining takes place
Castro et al. (1991)	1.42–8.14	Indian groups living close to mining sites in the Amazon

in the Santarem area. Mercury concentrations in hair of the riverine population ranged from 0.74–17.7 $\mu g\ g^{-1}$ of hair, and were up to five times higher that those measured in the samples from the Santarem population. Therefore, a direct relationship was established between mercury concentrations in fish and in humans.

Highest Hg concentrations in human hair were found paradoxically among riverine populations of the Negro River, central Amazon and among fishermen around the Tucuruí Reservoir, SE Amazon, where no gold mining occurs. Two factors may contribute to this result. First; the "blackwater" environment of the Negro River and the associated floodplain and limnological characteristics of artificial reservoirs favor high rates of methylation (Jackson 1986, 1988; Lacerda et al. 1989; Forsberg et al. 1994; Guimarães et al. 1994), and second, the riverine population from these areas consume up to 0.5 kg of fish daily. This is well above the ingestion rate believed to cause deleterious health effects, i. e., if the fish have more than 1.0 mg Hg g^{-1} (Aula et al. 1994). Since fish in both areas presented Hg concentrations at least twice as high (see Chap. 5), these results are not surprising.

Since fish is the major route of Hg to humans, it is logical to expect that one critical group with regard to Hg exposure would be Indians. A complete survey of people not directly involved with gold mining was done among the Yanomami Indians in Roraima State, where for the past 10 years prospectors have been invading and polluting the local environment with Hg. (Castro et al. 1991). These Indians are not directly involved in gold mining. Mercury concentrations in 162 individuals ranged from 1.42 – 8.14 $\mu g\ g^{-1}$ in females and from 1.40 – 7.90 $\mu g\ g^{-1}$ in males. The overall results showed that for certain groups over 40 % of the samples presented Hg concentration in hair higher than 6.0 $\mu g\ g^{-1}$. Methyl-mercury concentrations measured in these samples showed concentrations ranging from 0.96 – 3.02 $\mu g\ g^{-1}$, with a percentage variation ranging from 40 to 90 % of the total mercury concentrations. The average value was 3.61 $\mu g\ g^{-1}$, well below the established limit of 6.0 $\mu g\ g^{-1}$. Fortunately, the diet of the Yanomamis is not based on fish, but mainly animals, fruits and vegetables. Thus,this may be the reason for the low contamination found. Unfortunately, this is not the case for most other Amazonian Indian groups.

Barbosa et al. (1994 a, b), for example, studied the Hg concentrations in the hair of humans dwelling near the Fresco River mining site, a recently established operation inSE Amazon. They compared 745 samples from Kayapó Indians, who had no connection with gold mining, and prospectors, who were directly exposed to Hg occupationally, but with a very low intake of local fish. The results of that study (Table 6.6) showed that both groups were contaminated, but the Kayapó Indians, with a larger consumption of fish, presented much higher concentrations than prospectors. More than 59 % of blood samples from the Indians had Hg concentrations higher than 10.0 $\mu g\ l^{-1}$ and more than 23 % of hair samples had concentrations higher than 10.0 $mg\ g^{-1}$. On the other hand, only 33 % of the blood samples and only 6.2 % of the hair samples from the prospectors were higher than those concentrations. These results highlight the importance of environmental exposure as the major route of Hg to Indian groups in the Amazon, even when these groups are not involved with gold mining.

Another critical group exposed to environmental Hg are the riverine populations, which are relatively isolated and have extremely high fish consumption. Along the Madeira River, for example, dwellers from small villages, such as Humaitá and Manicoré, show higher Hg values in hair

Table 6.6. Total Hg content in human samples from Indian groups and prospectors from the Fresco River region, SE Amazon, Brazil. (After Barbosa et al. 1994)

Group	Hg in blood (% > 10.0 $\mu g\ l^{-1}$)	Hg in urine (% > 20.0 $\mu g\ l^{-1}$)	Hg in hair (% > 10.0 $\mu g\ g^{-1}$)
Indians	59 % (n = 132)	30 % (n = 194)	23.6 % (n = 419)
Prospectors	33 % (n = 129)	38 % (109)	6.2 % (n = 145)

than in Porto Velho, the capital of the state. This is mainly attributed to higher fish consumption rates in the more isolated, downstream villages, although similar Hg concentrations in the fish are found in the two areas. Porto Velho citizens, with better social and economical conditions, have a more diversified protein supply. The fish consumption rate in Porto Velho is less than two meals a week, while at Humaitá and Manicoré the rate is between three and four meals per week. Smaller villages will most likely have a much higher fish ingestion frequency, up to several meals a day (Malm 1992).

Barbosa et al. (1994a, b, 1995) and Boichio et al. (1995) also investigated the Hg level of 150 families of riverine population along the Madeira River, whose major diet is fish. Fish consumed by selected fishermen's families was up to 4.0 kg of fish per week. Most individuals (57%) showed a daily intake of Hg from consuming 40–200 g of fish, while 3% of the sampled families showed Hg intake from 200–1200 g of fish. Nearly 40% of the fish consumed by these families presented Hg concentrations between 0.3 and 0.5 $\mu g\ g^{-1}$, and 25% presented Hg concentrations higher than 0.5 $\mu g\ g^{-1}$.

Among 384 hair samples, 51% presented Hg concentrations higher than 10 $\mu g\ g^{-1}$; 11% higher than 30 $\mu g\ g^{-1}$ and 3% higher than 50 $\mu g\ g^{-1}$. From all samples, an average of 82% of the total Hg content was methyl-Hg. An interesting result of that study was that among children under 2 years old, who do no eat fish, 21% presented Hg concentrations higher than 10 $\mu g\ g^{-1}$. Since 53% of all women between the ages of 15 and 49 years also presented Hg concentrations in hair higher than 10 $\mu g\ g^{-1}$, the authors suggested placental transfer as a possible reason for the high Hg levels found in the young children.

The existing data on mercury contamination in humans in the Amazon gold mining region show that environmental exposure, in particular consumption of mercury-contaminated fish, is the major pathway of human contamination. Also, there seems to be already widespread contamination in the human populations of the Amazon region, including some Indian tribes. However, mercury poisoning itself has not been scientifically investigated and few conclusions can be drawn.

7 Perspectives on the Temporal Development of Mercury Inputs into the Environment

7.1 Global Inputs

Anthropogenic discharges of mercury are from electricity generation through coal and oil burning, solid and urban wastes, and from agriculture and forestry practices. These sources mobilize large amounts of Hg in the biosphere (ca. 11000 tons year^{-1}; Table 7.1), while release through natural weathering processes is less than one tenth of those from anthropogenic sources (ca. 900 tons year^{-1}). However, the mobilization of Hg from natural sources to the atmosphere of circa 2500 tons year^{-1} is, contrary to all other heavy metal pollutants, similar to the anthropogenic emissions to the atmosphere of nearly 3550 tons year^{-1} (Table 7.2). This natural atmospheric input is very high compared to all other heavy metals of environmental significance.

The high vapor pressure of Hg under normal surface conditions has led to a constant degassing of this metal from natural deposits and a global distribution throughout the atmosphere. Also, its origin in mantle fluids resulted in the presence of Hg as an impurity in many substances such as coal, oil, and minerals, in particular those associated with epithermal base-metal sulfides, resulting in extremely high volcanic emissions to the atmosphere (Mitra 1986). This characteristic is also responsible for the large atmospheric input from coal and oil burning for energy production, which constitutes the major indirect source of Hg from anthropogenic activities to the biosphere (Tables 7.1 and 7.2).

Table 7.1. Total global mercury inputs to the biosphere, excluding the atmosphere, in tons per year. (After Nriagu and Pacyna 1988; Nriagu 1989, 1990)

Agriculture and forestry	Urban waste	Electric generation, coal ash	Solid wastes weathering	Natural	Total
1960	1070	4400	2780	900	10310

Table 7.2. Present global mercury emissions to the atmosphere from natural and anthropogenic sources in tons per year. (After Nriagu and Pacyna 1988; Nriagu 1989, 1990)

Soil particles	Sea spray emissions	Volcanic	Forest fires	Biogenic sources	Total emissions
50	20	1000	20	1400	2490
Energy production	Ore refining and smelting	Waste incineration	Total anthro-pogenic sources	Total inputs	
2260	130	1160	3550	6040	

Another uniqueness of mercury is that while for most metals mining and refining of their ores are in general one of the largest anthropogenic sources of a given metal, this source is presently very small for Hg, resulting in total inputs to the environment of only 230 tons year^{-1}, which is much lower than natural volcanic emissions for example (Tables 7.1 and 7.2).

The importance of atmospheric inputs, from both natural and anthropogenic sources, accounting for over 50% of the sum of all other inputs, makes mercury contamination of regional and global significance rather than local, as with most other trace metals. Thus, any additional source of mercury to the environment may significantly affect its global biogeochemical cycle and may rapidly increase local and regional contamination levels. Atmospheric concentrations of Hg have increased ca. 1.5% in the northern hemisphere and ca. 1.2% in the southern hemisphere in the last 200 years. As a result, global deposition of Hg from the atmosphere may have increased by a factor of 0.5–5.1 since the begining of the 19th century (Slemr and Langer 1992).

7.2 The Decrease in Mercury Inputs from Industrial Point Sources

The seriousness of the situation has led to intense control of Hg emissions to the environment from industrial sources in most countries. This has resulted in a significant decrease in the global Hg inputs to the environment. For instance, estimates of anthropogenic Hg emissions to the atmosphere from the early 1970s generally gave values ranging from 10000 to 30000 tons year^{-1} (WHO 1976; Andren and Nriagu 1979). In the late 1980s, however, these estimates decreased, giving values from 1000–6000 to 13500 tons year^{-1} (Fitzgerald et al. 1984; Pacyna 1984; SCOPE 1985; Nriagu 1990), clearly showing the effectiveness of Hg control policies.

In Brazil, for example, where Hg utilization has been documented in detail, there has been a drastic decrease in Hg industrial uses and Hg fungicides have been banned, resulting in a strong decrease in Hg inputs to the environment. Thus, industrial and agricultural inputs to the environment in Brazil have presently been reduced to a few tons per year, two orders of magnitude lower than reported inputs for the late 1970s (Ferreira and Appel 1990, 1991). Chlor-alkali plants using Hg cells, Hg fungicides, and other organic Hg compounds previously used in agriculture are virtually disappearing in most countries. For example, Hg consumption in chlor-alkali plants in Brazil dropped from 80 tons year^{-1}in 1979 to only 18 tons year^{-1} in 1985, as a direct result of the substitution of Hg cells. This technology was responsible for 90% of chlorine production prior to 1980. Today, Hg-free technologies account for nearly 70% of alkali production in this country (Bezerra 1990).

More rigorouscontrol policies of such industries also resulted in considerably less emissions of Hg to the environment (Mitra 1986; Bezerra 1990). In Sweden, for example, prior to 1970, typically 50 to 85 g of Hg was lost to the environment for the production of 1.0 tons of chlorine. This emission factor decreased to less than 1 g ton^{-1} of chlorine in the late 1970s (Mitra 1986). Many new technologies to recover Hg from manufactured products, such as cells and fluorescent lamps, are presently being introduced to most industrialized countries, either as government-subsidized plants, such as in Japan, or as a private industry in a growing new market. The Dutch industry, for example, decreased Hg inputs to surface waters from 1700 kg year^{-1} in the 1970s to less than 500 kg year^{-1} in 1991 (Anonymous 1993).

As a direct result of the decrease in mercury inputs to the environment from anthropogenic sources, a significant decrease in mercury deposition rates and content in components of previously contaminated ecosystems occurred throughout the world. Lake sediments and fish are very good monitors of this decline.

In Finland, for example, after banning the use of phenyl-mercury as slimicide in the pulp and paper industry in the early 1970s, Hg inputs to freshwater ecosystems decreased, similarly to other European countries. Figure 7.1 shows the distribution profiles of Hg in sediment cores from lakes in Finland and Italy. Finnish lakes showed a decrease in their Hg content in superficial sediments by a factor of 100 (Lodenius 1991). In Lake Maggiore, Italy, theHg content decreased from 9.9 μg g^{-1} in the late 1970s to 1.1 μg g^{-1} in 1990 (Censi et al. 1991).

Mercury concentrations in freshwater fish of contaminated lakes also showed a rapid decrease. Average Hg concentrations in muscle tissue of pike (*Esox lucius* L.) dropped from 2.2 μg g^{-1} in 1973 to 1.4 μg g^{-1} in 1979, whereas in *Abramis faresus* L., they dropped from 0.87 μg g^{-1} to 0.43 μg g^{-1} during the same period. In both species, a constant annual reduction rate of 0.13 μg g^{-1} of Hg was determined (Nuorteva et al. 1979).

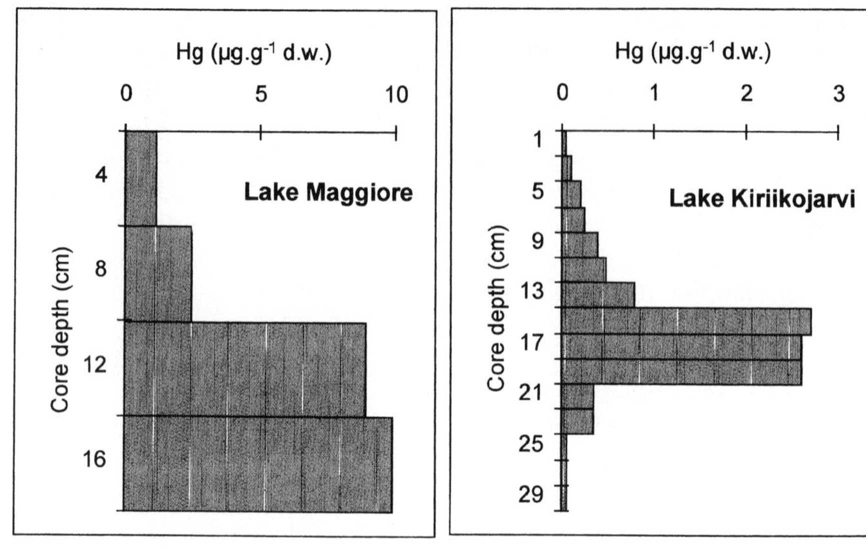

Fig. 7.1. Mercury distribution in sediment cores from contaminated lakes in Finland (Lodenius 1991) and Italy (Censi et al. 1991)

In China, Hg contents in fish from the contaminated Songhua River decreased 50 to 80% between 1975 and 1991. This resulted in a significant reduction in Hg intake by the local population of nearly 80% during the same period. Fisherman from the Songhua River showed a steady decrease in Hg concentrations in hair samples (Fig. 7.2), from an extreme value of 118 µg g^{-1} in 1976 to less than 20 µg g^{-1} in 1992 (Ming 1994).

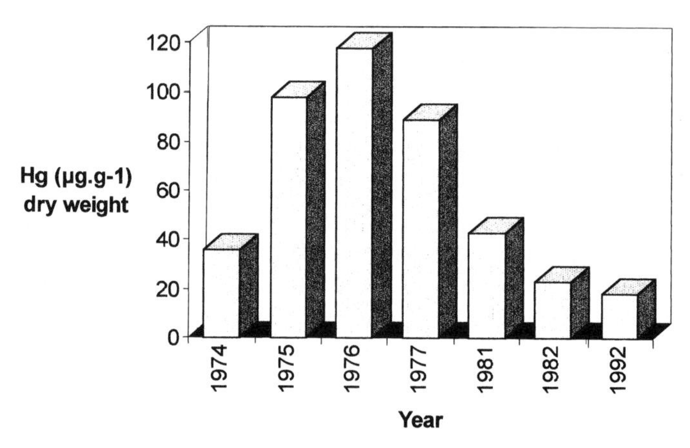

Fig. 7.2. Decrease in mercury concentrations (µg g^{-1}) in hair samples of fishermen from the Songhua River, China. (After Ming 1994)

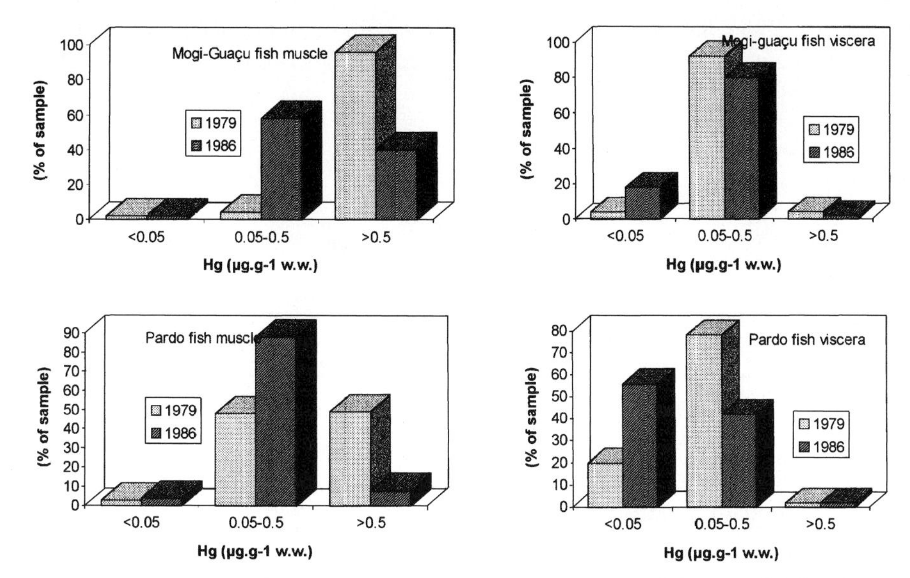

Fig. 7.3. Decrease in mercury concentrations in carnivorous fish sampled from two contaminated rivers in São Paulo State, Brazil, between 1979 and 1986. (Based on CETESB 1980, 1989; Boldrini et al. 1983; Boldrini 1990; Eysink 1990)

Other clear examples have been published recently from Brazil (Eysink 1990) and showed a significant decrease in mercury concentrations in sediments and fish from rivers and reservoirs previously submitted to industrial Hg inputs, attesting the efficiency of Hg control policies upon industrial Hg uses. Mercury concentrations in fish from contaminated rivers in São Paulo State, where major mercury pollution from industry occurs (CETESB 1980, 1989; Boldrini et al. 1983; Eysink 1990), were analyzed in two comparative surveys carried on in 1979 and 1986. A summary of their major results is presented in Fig. 7.3.

In 1979, from 84 carnivorous fish samples from the river Moji-Guaçu, 100% of muscle tissue samples showed Hg concentrations higher than 0.5 µg g^{-1} wet wt.; while in 1986, in the same river, only 39% of muscle tissue samples were higher than 0.5 µg g^{-1} wet wt. Mercury concentrations in viscera samples followed the same pattern. In the Pardo River, another heavily contaminated system, a survey from 1979 showed that half of the 50 fish muscle tissue samples analyzed had concentrations higher than 0.5 µg g^{-1}, while in 1986 no muscle tissue sample showed a concentration higher than that value. In viscera, 96% of samples showed Hg concentrations between 0.05 and 0.5 µg g^{-1} wet wt. in 1979, this value dropped to less than 50% in 1986. Other fish samples from herbivorous, omnivorous, and detritivorous species showed the same behavior.

Table 7.3. Reduction in mercury concentrations (mean and range) in selected fish species (μg g^{-1} wet wt.) from Lake St. Clair, Canada, between 1970 and 1980. (After Ogilvie 1981)

Fish species	1970	1980	Reduction (%)
Cyprinus carpio (carp)	0.92 (0.38–2.20)	0.57 (0.04–1.20)	38.0
Esox lucius (pike)	2.87 (0.61–9.90)	1.09 (0.20–2.80)	62.0
Morone chrysops (bass)	1,99 (0.47–5.70)	0.49 (0.14–1.40)	75.4
Ictaluras punctatus (catfish)	1.20 (0.33–3.00)	0.78 (0.18–1.90)	35.0
Perca flavescens (perch)	1.48 (0.02–8.20)	0.29 (0.10–0.99)	80.4

In Lake St. Clair, Ontario, Canada, efficient Hg emission control policies also resulted in a reduction of the Hg content in fish. Chlor-alkali plants located 40 km upstream of Lake St. Clair have been operating since 1950 with emissions of 31 kg Hg day^{-1}, resulting in extreme contamination. Annual catches of over 400 tons of fish ceased in the lake when concentrations over 10.0 μg Hg g^{-1} were reported in many fishes of economic importance. After government pressure, emission from these factories decreased to 0.4 kg day^{-1} within a few months in 1970. In 1973, the old plants were closed and replaced by Hg-free technologies for chlorine production. The fish responded promptly to these measures (Table 7.3). The percent reduction of Hg concentrations ranged between 35% for catfish to 80.4% for yellow perch, thus allowing commercial fisheries to start again in the lake.

The results discussed above clearly show that the control of Hg emissions from industrial sources in both the Western World as well as in the Third World countries was quite efficient and resulted in a significant decrease in Hg concentrations in the environment. New, more restrictive legislation regarding Hg uses in industry and agriculture is being put forward in many countries. Solid wastes containing Hg and manufactured refuses from diverse industrial processes, such as used fluorescent lamps and batteries, are being recycled throughout the world and decreased Hg consumption in industry is evident in many countries. New technologies that are more efficient and cheaper for most classical Hg uses such as in the pesticide or chlorine industry are presently available. Therefore, it is reasonable to expect a further reduction in environmental Hg levels in previously contaminated sites, at least when agriculture and industrial processes are the major source of this element. "Re-birth" of Hg contamination, however, may still occur when contaminated "sinks", such as aquatic sediments, are reworked by dredging and other civil engineering works.

7.3 Temporal Development in Gold-Mining Regions: The Second Gold Rush and Before

The temporal development of mercury inputs into the environment in Brazil has been studied through the analysis of cores from lakes. These lakes were close to mining sites in order to record the near-field effects (Fig. 7.4) and from remote mountainous areas to record the far-field effects (Fig. 7.5).

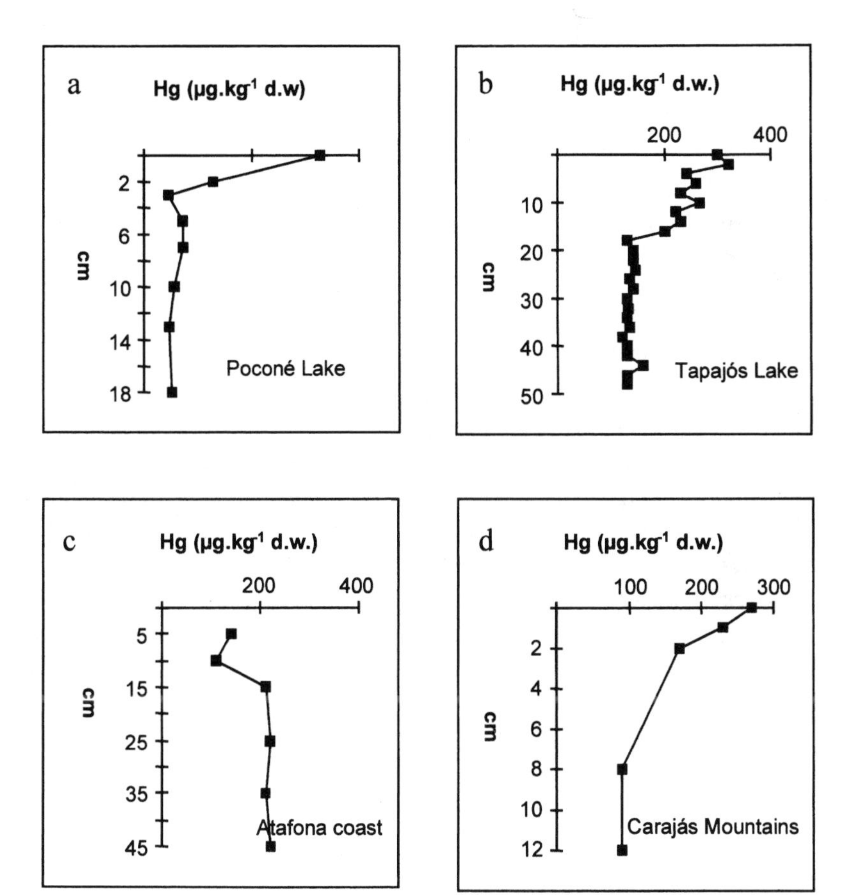

Fig. 7.4. Mercury (µg kg^{-1}) in sediment cores (depth in cm) from remote lakes in the Brazilian areas affected by gold mining. Only relatively remote areas outside the direct influence of mining operations were reported. **a** Pantanal Lakes, central Brazil (Lacerda et al. 1991 c); **b** Tapajós Still Waters, SE Amazon (Müller 1993); **c** Paraíba do Sul River Delta, Atafona, SE Brazil (Lacerda et al. 1993); **d** Carajás lakes SE Amazon. (Lacerda, unpubl.)

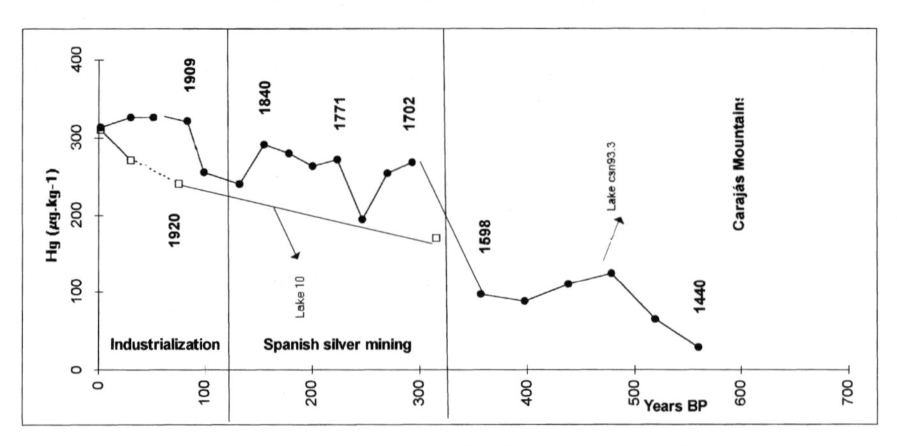

Fig. 7.5. Mercury concentrations along a dated core from the the Carajás plateau, S Amazon. (Lacerda, unpubl. data)

Mercury concentrations were higher in surface sediments of all areas (Fig. 7.5), ranging from approximately 60 µg kg^{-1} in lakes in the Pantanal area to values as high as 700 µg kg^{-1} in the still waters of the Tapajós River. Background Hg concentrations, considered here as the Hg concentration when the profile becomes constant with depth, also varied among sites. In lakes of the Pantanal formation, a Quaternary sandy plain, background Hg concentrations were between 10 and 30 µg kg^{-1}, whereas in the Tapajós area background concentrations were from 80 to 100 µg kg^{-1}. In the Carajás lakes, with no fluvial inputs of sediments, organic matter reaches over 95 % of the sediment. Here, background Hg concentrations can reach up to 170 µg g^{-1}.

All the Hg concentrations found are lower when compared to freshwater sediment cores from the northern hemisphere. Evans (1986) reported background values for remote Canadian lakes to be circa 100 µg kg^{-1}, while surface sediments presented values up to 560 µg kg^{-1}. Andren and Nriagu (1979) considered background values for northern hemisphere lakes of circa 330 µg kg^{-1}. Rekolainen et al. (1986) found that background Hg concentrations in remote lakes of Finland ranged between 50 to 250 µg $kg^{-1,}$ but found values as high as 550 µg kg^{-1} in the surface layers of these lakes. These data are considerably higher than the concentrations found in the freshwater system affected by gold mining in the Amazon region and are due to the presence of Hg-rich base rocks in most Canadian and north European areas, thus reflecting the large anthropogenic Hg emissions from industry in these areas.

Notwithstanding the relatively low concentrations, it is clear that Hg contamination is significant. Moreover, it is a regional rather than a local phenomenon, at least in the Amazon region. The rather short cores with a sedimentation history of only this century clearly show the impact of recent gold-mining activities. Information on the historical emissions of mercury from the colonial period were obtained from dated cores from the the Carajás plateau, S. Amazon

Based on historical date (see Chap. 1), three major sources of Hg to the atmosphere of South America were active over the last 500 years. From 1570 to the end of the 19th century, extensive silver mining using Hg amalgamation resulted in an estimated emission of ca. 200 000 tons of Hg, with yearly inputs ranging from 400–1100 tons (Nriagu 1994). A second probable source would be the Hg released into the northern hemisphere following the industrial revolution, although it has never been considered as capable of significantly affecting trace metal deposition over southern latitudes. Industrialization in the southern hemisphere is a very recent phenomenon of minor importance and difficult to detect in remote areas, in particular in the lakes studied here having a slow accretion rate. The last Hg source for the region is the present gold rush in Amazon countries which started in the late 1970s and has been well documented.

Figure 7.5 shows the Hg concentrations through the sediment core of Lake CSN93, in the Carajás plateau, S. Amazon. Sediment is mostly composed of organic matter (% Org-C >80%.) and shows very slow and constant deposition rates (<0.02 cm year^{-1}). Since the lake has a very small basin and no fluvial input, this deposition is mostly from the atmosphere. The constancy of major sediment parameters with depth indicates no bioturbation or remobilization of core layers (Turq, pers. comm.), allowing a good assessment of Hg deposition over time.

The top 5.0 cm of the cores roughly corresponds to the last 150 years, when major industrialization occurred in the northern hemisphere. Mercury concentrations in lakes increased from 255 µg kg^{-1} around 1880 A.D. to 320 µg kg^{-1} in 1994 A.D. in lake CSN93.3. The lake shows a consistent increase in Hg concentrations over the past 150 years. Since most industrialization in the south is very recent, 30 years old in most countries, and the Amazon gold rush only started in the 1970s, any increase in Hg deposition due to these sources would be restricted to the top 1.0 cm of the cores. Therefore, these results suggest that industrialization in the north did affected Hg deposition over the southern hemisphere.

A comparison of Hg distribution in Lake CSN93.3 with the estimated Hg emissions from the Spanish silver mines (Fig. 2.2) shows that the two curves match fairly well. Mercury concentrations in the profile increase from <100 µg g^{-1} before 1580 A.D. to nearly 300 µg g^{-1} from 1750 to 1850, when the largest annual emissions occurred. Although results from a single core should be taken as preliminary, we conclude that the variations observed are due to inputs from colonial silver mining since no other mercury source for that period can be identified.

The evidence from the cores shows that the mercury from gold mining activities, both in the past and more recently, has a regional distribution and effects areas far from where the actual extraction has taken place.

8 Summary and Outlook

It is difficult to draw parallels between mercury contamination in temperate climates, where most of the research has been carried out, and the tropical climates. Differences exist between the use of mercury (socio-economic dimension) and its cycling in the environment (ecological dimension).

The input of mercury by industrial point sources into the environment has decreased drastically, in both developed and developing countries. The remaining issues, in particular in temperate climates, are the accumulated levels of mercury in hot spots like industrial sites and contaminated soils and sediments. The appropriate technology for cleanup of hot spots is in principle available (Ebinghaus et al. 1997). The mercury, which is spread throughout the environment and has accumulated in soils through atmospheric deposition or in sediments on river floodplains, lakes and coastal areas, will be a potential danger for aquatic and terrestrial ecosystems. For sediments, it can be expected that they will, through natural processes, be covered with new uncontaminated deposits and, hence, isolate the contaminated layers from the aquatic ecosystem. This will not be the case for soils; here, we only have the natural process of degassing. Hence, it can be expected that contaminated soils will be a nearly permanent source of mercury in the ecosystem. Furthermore, changing environmental conditions (changing land use, acidification, etc.) which cause changes in the capacity-controlling parameters (Salomons and Stigliani 1995), may enhance or reduce current impacts.

The issue in developing countries is quite different. Although emissions for point sources have decreased here also, the "diffuse" use of mercury in gold mining has increased over the past years and is likely to increase in the future. Its attractiveness lies in being a reliable and cheap technology easy to operate at the individual scale. Furthermore, mercury is used by many groups of individuals in widely dispersed and often remote areas. Controlling this type of use is more difficult compared with controlling mercury discharges from (few) industrial point sources. Furthermore, this type of pollution abatement has a strong socio-economic dimension. Therefore, notwithstanding the recent regulations making the acquisition and use of Hg in many countries more difficult, this type of technology will only be limited by the exhaustion of gold and silver deposits or by a drastic decrease in gold prices. Both alternatives are unlikely to occur in the next 50 years, and new prospecting areas are expected to be opened, in particular in northern South America and Southeast Asia.

Research in tropical climates, in particular the results from Brazil, make it possible to schematize the dispersion of mercury and its transformation to the more toxic methyl-mercury species (Fig. 8.1).

Excess liquid mercury from the amalgamation process remains at the site of extraction and can be found in "hot spots" in the main tailing areas or on the riverbed. Mercury in the "hot spots" is not biologically available because it is in its metallic form. Moreover, in the tailings and in the fast flowing large rivers of the Amazon, the conditions for transformation to methyl-mercury are not favorable. However, given enough time, the mercury from these areas will become dispersed into the environment through erosion and leaching. Much more important, however, is the gaseous mercury released during the extraction process. Evidence from dated cores from lakes showed the high dispersion in the environment of these emissions. Equally important is the transformation of gaseous mercury to inorganic mercury ions through atmospheric processes.

Tropical forest areas characteristically present extensively flooded areas. For example, the Brazilian Amazon has over 200 000 km^2 of seasonally flooded forests. These areas are covered by 6 to 15 m of water during 6 or more months every year. During these months, most fish migrate to flooded areas for breeding and feeding (Ayres 1994). Flooded soils meet all

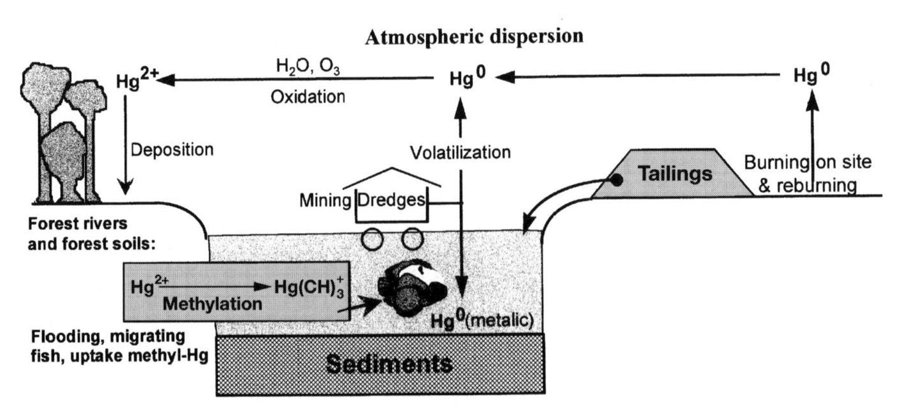

Fig. 8.1. Hypothetical link between atmospheric Hg and Hg content in fish from gold mining areas in the tropics

the necessary conditions for high methylation rates. Thus, it is expected that the seasonal migration of fish will result in increasing concentrations of methyl-Hg. Eventually, methyl-Hg will also be exported to major rivers during the lowering of the water level. This, rather than direct inputs of Hg to rivers and soils, is probably responsible for the elevated Hg levels found in most fish, even those collected far from any mining site.

The long-term cycling of elements in the tropical environment shows large differences within temperate climate areas. These differences have to be taken into account when making predictions of future impacts, i.e., once the mining activities have ceased but the mercury is still present in the ecosystem. The cycling of inorganic substances in the Amazon forest is very efficient. Biological interactions at the root–soil (mycorrhiza) and atmosphere–canopy (epiphylae and high biomass and diversity of epiphytes) levels are extremely efficient for the uptake of nutrients from atmospheric precipitation, allowing quickincorporation into the plant biomass. In temperate forests, this pathway of nutrients is negligible. In temperate ecosystems, the larger nutrient reservoir is the soil, whereas in tropical forest it is the plant biomass itself. Even when these nutrients are eventually returned to the soil as plant litter, fast decomposition rates mediated by root-associated fungi recycle them very fast, avoiding large losses to subsoil and loss from the ecosystem.

The major pathway for mercury in the Amazon is the atmosphere. Mercury vapor is readily absorbed from the atmosphere by plant leaves, therefore, it is likely that it will follow the same pathway as nutrients. This is a major difference compared with the fate of mercury in temperate ecosystems, in which immobilization in soils and other slow cycling compartments dominate. In most temperate ecosystems when sources of a given contaminant are controlled, one can expect that over time it will eventually accumulate in long-term sinks, and that concentrations in cycling compartments will constantly decrease unless the capacity of the sink is exceeded or external changes occur (time-delayed responses).

Through efficient recycling in Amazon ecosystems, mercury (and other contaminants) just moves from one cycling, biological compartment to another, increasing the probability of organification and accumulation in high trophic level animals. Accumulation of pollutants in the biomass (tropical system) instead of sinks like soils (temperate systems) makes it much more difficult to manage the system once pollutants have entered it.

Biological diversity in most tropical ecosystems is strongly dependent on the structure of food webs, which are in general controlled by toppredators. It is exactly in these animals that mercury should accumulate, and where its toxic effects should appear first. Therefore, once high trophic level species are affected, whole food chains can also be affected, resulting in rapid changes in community structures. It is likely that fish communities would be one to be highly affected. Fish, apart from the forest biomass itself, is the major natural resource in the Amazon for the local population. Therefore, any changes in such important resources may lead to economic constraints in the region.

The picture presented on the fate of mercury in a tropical system needs more validation and more research. However, the important differences within temperate climates are the high turnover and the biomass as the main reservoir of nutrients (and of pollutants). This fact makes it very difficult to manage the system when pollutants enter its intricate element cycles. Once they have entered it, it will be more difficult to control them compared with the temperate climate ecosystems. The question whether mercury (and other pollutants) in the Amazon or in any other tropical ecosystem presents a potential „chemical time bomb“ with its associated time-delayed and spatially displaced responses has to be answered in the positive sense. However, two remarks have to be made to be put this conculsion into the context of the more detailed experience from temperate climates:

- The "chemical time bomb" in a tropical ecosystem will be much more difficult to control due to the predominant role of the biomass as a sink.
- The existence of a "chemical time bomb" and its associated delayed effects has to be substantiated with more research; not with isolated studies nor with multidisciplinary research, but with integrated research with the functioning of the ecosystem as its focal point.

However, some practical recommendations to restrict emissions and with regard to handling existing problems can be made. With regard to exposure to methyl-Hg, it became clear that the critical populations are riverine populations that routinely eat fish and not the goldminers, who have better economic conditions and diversified food options. Regarding Hg^{0} inhalation, gold dealers in indoor shops are critical groups rather than garimpeiros. These two critical groups should receive special attention regarding exposure risks.

Due to the high Hg concentrations found in fish and hair samples from riverine populations, a preliminary recommendation for critical groups (women of child-bearing age and children) is to avoid routine ingestion of carnivorous fish.

In critical environments like gold dealer shops, it is advisable to use open rooms with adequateventilation and efficient treatment/exhaustion systems for Hg vapor retention after reburning, in order to avoid mercury poisoning. In garimperos, the use of retorts is strongly recommended since they can drastically reduce Hg emissions.

References

Achmadi UF (1994) Occupational exposure to mercury at the gold mining: a case study from Indonesia. In: Evironmental mercury pollution and its health effects in Amazon River Basin. Natl Inst Minamata Disease and Inst Biophysics of the Univ Federal do Rio de Janeiro Rio de Janeiro, pp 10–16

Akagi H, Mortimer DC, Miller DR (1979) Mercury methylation and partition in aquatic systems. Bull Environm Contam Toxicol 23:372–376

Akagi H, Kinjo Y, Branches F, Malm O, Harada M, Pfeiffer WC, Kato H (1994) Methylmercury pollution in Tapajós River Basin, Amazon. Environ Sci 3:25–32

Akagi H, Malm O, Branches FJP, Kinjo Y, Kashima Y, Guimarães JRD, Oliveira RB, Haraguchi K, Pfeiffer WC, Takizawa Y, Kato H (1995) Human exposure to mercury due to gold mining in the Tapajós River Basin Amazon Brazil: speciation of mercury in human hair, blood and urine. Water Air Soil Pollut 80:85–94

Alberts JJ, Schindler JE, Miller RW (1974) Elemental mercury evolution mediated by humic acid. Science 184:895–897

Alvarez-León R (1994) A review of seventeen years of speciation and research on the role of mercury in the north of Colombia. In: Evironmental mercury pollution and its health effects in Amazon River Basin. Natl Inst Minamata Disease and Inst Biophysics of the Univ Federal do Rio de Janeiro Rio de Janeiro, pp 21–22

Andrade JC, Bueno MIMS, Soares PV, Choudhuri A (1988) The fate of mercury released from prospecting areas (Garimpos) near Guarinus and Pilar, Goiás (Brazil). An Acad Brasil Ciênc 60:293–303

Andren AW, Nriagu JO (1979) The global cycle of mercury. In: Nriagu JO (ed) The biogeochemistry of mercury in the environment. Elsevier, Amsterdam, pp 1–22

Anonymous (1993) Process recover resuable mercury. Am Met Mark 101(2):7

Araujo Neto H (1990) Poluição causada pelo mercúrio no garimpo de Serra Pelada, Pará. In: Hacon S, Lacerda LD, Carvalho D, Pfeiffer WC (eds) Riscos e Consequências do Uso do Mercúrio, FINEP/UFRJ, Rio de Janeiro, pp 263–267

Artaxo PF, Gerab ML, Rabello C (1993) Elemental composition of aerosol particles from two background monitoring stations in the Amazon Basin. Nuclear Instru Methods Phys Res B75:277

Aula I, Braunsweiller H, Leino T, Malin I, Porvari P, Hatanaka T, Lodenius M, Juras A (1994) Levels of mercury in the Tucuruí reservoir and its surrounding area in Pará Brazil. In: Watras CJ, Huckabee JW (eds) Mercury pollution: integration and synthesis. Lewis, Boca Raton, pp 21–40

Averill CV (1946) Placer mining for Gold in California. California Division of Mines Bulletin no 135, Sacramento, California, p 377

Ayres JM (1994) As Florestas do Mamirauá. CNPq/Soc Civil Mamirauá, Belém

Bacci E, Focardi S, Leonzio C, Renzoni A (1980) Mercury concentration in muscle, liver and stomach content of *Mullus barbatus* of the northern Tyrrhenian. Metals Mediterr Newslett 2:5–10
Baeyems W, Leermakers M, Dedeuwaerder H, Larsens P (1991) Modelization of the mercury fluxes at the air sea interface. Water Air Soil Pollut 56:731–744
Bakir F, Damluji SF, Amin-Zaki L, Murdatha M, Khelidi A, Al-Ranei NY, Tikriti S, Dhahir HI, Clarkson TW, Smith JC, Doharty RA (1973) Methyl-mercury poisoning in Iraq. Science 181:230
Barbosa AC, Boischio AAP, East GA, Ferrari I, Golçalves A, Silva PRM, Cruz TME (1995) Mercury contamination in the Brazilian Amazon. Environmental and occupational aspects. Water Air Soil Pollut 80:109–121
Bargagli R (1995) Effects of abandoned mercury mines on terrestrial and aquatic ecosystems. In: Proc 3rd Int Conf Biogeochemistry of trace elements, Paris Abstr B1
Benning D (1958) Outbreak of mercury poisoning in Ohio. Ind Med Surg 27:354–363
Benzing DH (1981) Mineral nutrition of epiphytes: an appraisal of adaptative features. Selbyana 5:219–223
Bezerra JFM (1990) Estimativas de cargas de mercúrio para o meio ambiente por atividades industriais. In: Hacon S, Lacerda LD, Carvalho D, Pfeiffer WC (eds) Riscos e Consequências do Uso do Mercúrio, FINEP/UFRJ, Rio de Janeiro, pp 91–109
Bidstrup PL, Bonnell JA, Harvey DG, Locket S (1951) Chronic mercury poisoning in men repairing direct-current meters. Lancet 2:856–861
Bisogni JJ (1979) Kinetics of methylmercury formation and decomposition in aquatic environments. In: Nriagu JO (ed) The biogeochemistry of mercury in the environment., Elsevier, Amsterdam, pp 211–227
Bjorklund I, Borg H, Johansson K (1984) Mercury in Swedish lakes – its regional distribution and causes. Ambio 13:118–121
Bloom NS, Fitzgerald WF (1988) Determination of volatile mercury species at the picogram level by low-temperature gas chromatography with cold vapour atomic fluorescence. Anal Chim Acta 208:151–161
Bloom NS, Porcella DB (1994) Less mercury. Nature 367:694
Bloom NS, Watras CJ, Hurley JP (1991) Impact of the acidification on the methylmercury cycle in remote seapage lakes. Water Air Soil Pollut 56:477–491
Boischio AAP, Henshel D, Barbosa AC (1995) Mercury exposure through fish consumption by the upper Madeira River population, Brazil – (1991). Ecosyst Health 1:177–192
Boldrini CV (1991) Mercúrio na Baixada Santista. In: Hacon S, Lacerda LD, Pfeiffer WC, Carvalho D (eds) Riscos e Consequências do Uso do Mercúrio. FINEP/UFRJ, Rio De Janeiro, pp 161–195
Boldrini CV, Padua HB, Pereira DN, Rezende EK, Juras AA (1983) Contaminação por mercúrio nos rios Moji-guaçú e Pardo. Rev DAE 135:106–117
Brabo EE (1992) Contaminação por Mercúrio dos Rios Crepori e Marupã Bacia do Tapajós Pará T MSc, Ctr Geociências, Univ Fed Pará, Belém
Brabo ES, Costa MQ, Ramos JFF (1991) Avaliação preliminar da contaminação por mercúrio nos rios Creparí e Marupá. In: Proc III Simp Geol Amazônia 1:331–336
Brading DA, Cross HE (1972) Silver mines of Colonial America. Hispanic Am Hist Rev 52:547–549
Branches FJP, Erickson T, Aks SE, Hryhorczuk DO (1993) The price of gold: mercury exposure in the Amazonian Rain Forest. J Clin Toxicol 31:295–306

Branches FJP, Malm O, Bastos WR, Pfeiffer WC (1992) Clinical findings, its relation with air mercury concentrations and urinary Hg levels among gold shop workers, Amazon, Brazil. Abstr 4th Annu Meet Int Soc Environ Epidemiol, Mexico Session: 03.1

Brasil (1975) Resolução No 18/75 da Comissão Nacional de Normas e Padrões para Alimentos Resolução de 18 de agosto de 1975. Ministério da Saude. Diário oficial da União, Brasília, 19 de Dezembro de 1975. Seção 1 p 16378

Brosset C, Lord E (1991) Mercury in precipitation and ambient air. Water Air Soil Pollut 56: 493-506

Browne CL, Fang SC (1978) Uptake of mercury vapor by wheat: an assimilation model. Plant Physiol 61: 430-433

Brüseke FJ (1993) Mineração, ouro e a caotização de uma região. In: Mathis A, Dhaag R (eds) Consequências da Garimpagem no Âmbito Social e Ambiental da Amazônia. Buntstiff e.v., Katalise, FASE, Belém, pp 38-45

Bycroft BM, Coller BAW, Deacon GB, Coleman DJ, Lake PS (1982) Mercury contamination of the Lerderderg River, Victoria, Australia, from an abandoned gold field. Environ Pollut Ser A 28: 135-147

CACEX (1988) Anuário Mineral. Carteira de Comércio Exterior do Branco do Brasil, Rio de Janeiro

Calazans CF, Malm O, Pfeiffer WC (1993) Evaluation of atmospheric Hg concentrations through the biological monitor *Tillandsia usneoides* (Bromeliaceae). In: Proc Int Conf Heavy metals in the environment, Toronto 1: 344-347

Callaham JE, Miller JW, Craig JR (1994) Mercury pollution as a result of gold extration in North Carolina, USA. Appl Geochem 9: 235-241

Câmara VM (1985) Estudo Comparativo dos Efeitos Tardios dos Fungicidas Organomercuriais no Munícipio de Campos, RJ. T Dout, Escola Nac Saúde Pública, ENSP/FIOCRUZ, Rio de Janeiro

Câmara VM (1994) Epidemiological assessment of the environmental pollution by mercury due to gold mining in the Amazon region. In: Evironmental mercury pollution and its health effects in Amazon River Basin. Natl Inst Minamata Disease and Inst Biophysics of the Univ Federal do Rio de Janeiro Rio de Janeiro, pp 80-84

Camara VM (1986) Teores de mercúrio no cabelo: um estudo comparativo em trabalhadores da lavoura de cana de açúcar com exposição pregressa aos fungicidas organo-mercuriais no municipio de Campos-RJ. Cad Saude Pública 2: 359-372

Castro MB, Albert B, Pfeiffer WC (1991) Mercury levels in Yanomami indians hair from Roraima, Brazil. Proc 8th Int Conf Heavy metals in the environment, Edinburgh 1: 367-370

Censi R, Baudo R, Muntau H (1991) Mercury deposition history of Pallanza Bay, Lake Maggiore, Italy. Environ Technol 12: 705-712

CETEM (1989) Relatório Anual do Projeto Poconé. Centr Tecnol Mineral, Rio de Janeiro, p 287

CETESB (1980) Avaliação da Contaminação Atual dos Rios Moji-guaçú e Pardo e seus Reflexos sobre Comunidades Biológicas. Comp Est Tecnol San Bas, São Paulo, 3 vols, p 637

CETESB (1989) Avaliação do Quadro de Contaminação por Mercúrio nos Rios Pardo e Moji-guaçú (SP). Comp Est Tecnol San Bas, São Paulo, p 126

Ching L, Hongxiao T (1985) Chemical studies of aquatic pollution by heavy metals in China. In: Irgolic KJ, Martell AE (eds) Proc Worksh Environ Inorganic Chem, San Miniato, FL, pp 359-371

CIMELCO (1991) Proyecto Binacional Puyango-Tumbes. Consorcio Cimelco Construtores, Tumbes, Peru

Clarkson TW, Hamada R, Amin-Zaki L (1984) Mercury. In: Changing metal cycles and human health. Springer, Berlin Heidelberg New York, pp 285–309

Cleary D (1990) Anatomy of the Amazon Gold Rush. Macmillan, London, p 245

Cleary D (1994) Mercury contamination in the Brazilian Amazon: an overview of epidemiological studies. In: Evironmental mercury pollution and its health effects in Amazon River Basin. Natl Inst Minamata Disease and Inst Biophysics of the Univ Federal do Rio de Janeiro Rio de Janeiro, pp 61–72

Cleary D, Thornton I, Brown N, Kenzatis G, Delves T, Worthington W (1994) Mercury in Brazil. Nature 369:613–614

CODECHOCO (1991) Estudio del Impacto Ambiental en Zonas Mineras de los Municípios de Taolo, Condoto y Istmina. I. Parte final. Corporación Nacional para el Desarrollo del Choco, Bogota, p 16

Compeau GC, Bartha R (1985) Sulfate reducer bacteria: principal methylators of mercury in anoxic estuarine sediments. Appl Environ Microbiol 50:498–452

Copplestone JF, McArthur DA (1967) An inorganic mercury hazard in the manufacture of artificial jewellery. Br J Ind Med 24:77–80

Cossa D, Martin JM (1991) Mercury in the Rhône delta and adjacent marine areas. Mar Chem 36:291–302

Cowgill UM (1975) Mercury contamination in a 54-m core from lake Huleh. Nature 256:476–478

Cramer SW (1990) Problems facing the Philippines. Inter Mining July 1990:29–31

Crowder A, St-Cyr L (1991) Iron oxide plaques on wetland roots. Trends Soil Sci 1:315–329

DePaula FC (1989) Geoquímica de Sedimentos da Bacia Ocidental do Rio Madeira, Rondônia. MSc Thesis, Univ Fed Fluminense, Niteroi, p 71

DePaula FCF, Lacerda LD (1991) Geoquímica de sedimentos de fundo da bacia do Rio Madeira, Rondônia e sua relação com a classe de rios amazônicos. Ann 3rd Cong Geol Amazônia Belém 1:527–540

D'Itri FM (1992) Mercury pollution and cycling in aquatic systems. In: Vermet J-P (ed) Proc 5th Int Conf Environ Contamination, CEP Consultants, Edinburgh, pp 64–74

D'Itri PA, D'Itri FM (1977) Mercury contamination: a human tragedy. Wiley, New York, p 311

DNPM (1983) Garimpos do Brasil Publ Avulsa no 5, Dept Nac Producao Mineral, Brasilia, DF, p 378

DNPM (1988) Panorama do setor mineral do Pará. Dept Nac Produção Mineral, V Distrito, Belém, p 41

DNPM (1989) Anuario Mineral Brasileiro. Dept Nacional da Produção Mineral, Brasília, DF

DNPM (1992) Anuario Mineral Brasileiro. Dept Nacional da Produção Mineral, Brasília, DF

Driscoll JN (1974) Sampling and analytical techniques for mercury in stationary sources: a state-of-the-art report. Health Lab Sci 11:348–353

Drummond RA, Olson GF, Batterman AR (1974) Cough response and uptake of mercury by brook trout *Salvelinus fontinalis*, exposed to mercuric compounds at different hydrogen ion concentrations. Trans Am Fish Soc 103:244–249

EPA (1971) Report on the pollution affecting water quality of the Cheyenne River system, western South Dakota. EPA, National Field Investigations Center, Denver, Colorado

Ermanov VV (1995) Biogeochemical food chains of the south Fergana subregion of the biosphere and their correction. In: Proc 3rd Int Conf Biogeochemistry of trace elements, Paris Abstr B1
Ernst WHO (1988) Response of plants and vegetation to mine tailings and dredge materials. In: Salomons W, Forstner (eds) Chemistry and biology of solid wastes. Springer, Berlin Heidelberg New York, p 305
Evans RD (1986) Sources of mercury contamination in the sediments of small headwater lakes in South-Central Ontario, Canada. Arch Environ Contamin Toxicol 15:505–512
Eysink GGJ (1990) A presença de mercúrio nos ecossistemas aquáticos do Estado de São Paulo. In: Hacon S, Lacerda LD, Pfeiffer WC, Carvalho D (eds) Riscos e Conseqüências do Uso de Mercúrio. FINEP/UFRJ Rio de Janeiro, pp 12–28
Farid LH (org) (1992) Diagnóstico preliminar dos impactos ambientais gerados por garimpo de ouro em Alta Floresta, MT. Ser Tecnol Ambiental CETEM, 2:1–190
Farid LH, Machado JEB, Silva AO (1991) Emission control and mercury recovery from mining (garimpo) tailings: Poconé experience. In: pp 217–224 Proc Int Symp Environ Studies on Tropical Rain Forests, Biosfera Ed, Rio de Janeiro
Fearnside PM (1994) Desmatamento na Amazônia
Fernandes FRC, Portella ICMHM (1991) Recursos Minerais da Amazônia: Alguns Dados Sobre Situação e Perspectivas. Estudos e Documentos no 14. Centro de Tecnologia Mineral (CETEM-CNPq), Rio de Janeiro, p 44
Fernandes RS, Guimarães AF, Bidone ED (1990b) Monitoramento do mercurio na area do Projeto Carajás. Saneamento Ambiental 6:34–41
Fernandes RS, Guimarães AF, Bidone ED (1990a) Monitoramento do mercúrio na area do Projeto Carajás. Bios 2:37–44
Ferrara F, Petrodino A, Maserti E Seritti A, Barghigiani C (1982) The biogeochemical cycle of mercury in the Mediterranean. Part II. Mercury in the atmosphere, aerosol and in rain of a northern Tyrrhenian area. Environ Technol Lett 3:449–456
Ferrara R, Maserti BE (1992) Mercury concentration in the water, particulate matter, plankton and sediment of the Adriatic Sea. Mar Chem 38:237–249
Ferrara R, Maserti BE (1994) Mercury degassing rate in some mediterranean areas. In: Proc Int Conf Mercury as a global pollutant, Whistler, BC, Abstr Sec 8B
Ferreira JR, Devol AH, Martinelli LA, Forsberg BR, Victoria RL, Richey JE, Mortatti J (1988) Chemical composition of the Madeira River: seasonal trends and total transport. Mitt Geol Palant Inst Univ Hamburg 66:63–76
Ferreira RCH, Appel LE (1990) Estudo detalhado de fontes e usos de mercúrio. Relatório Preliminar. Centro de Tecnologia Mineral/Conselho Nacional de Desenvolvimento Científico e Tecnolólico, CETEM/CNPQ, Rio de Janeiro, p 62
Ferreira RCH, Appel LE (1991) Mercury: sources and uses in Brazil. Ann I Int Symp on Environmental studies on tropical rain forests, Manaus, pp 207–216
Fisher JR (1977) Silver mines and silver miners in colonial Peru, 1776–1824. Centre for Latin American Studies, Univ Liverpool
Fitzgerald WF, Gill GA, Kim JP (1984) An equatorial Pacific Ocean source of atmospheric mercury. Science 224:597–599
Fitzgerald WF, Mason RP, Vandal GM (1991) Atmospheric cycling and air-water exchange of mercury over mid-continental lacustrine regions. Water Air Soil Pollut 56:745–767
Florentine MJ, Sanfilippo DJ (1991) Elemental mercury poisoning. Clin Pharm 10:213–221

Forsberg BR, Forsberg MCS, Padovani CR, Sargentini E, Malm O (1994) High levels of mercury in fish and human hair from the Rio Negro basin (Brazilian Amazon): natural background or anthropogenic. In: Environmental mercury pollution and its health effects in Amazon River Basin. Natl Inst Minamata Disease and Inst Biophysics of the Univ Federal do Rio de Janeiro Rio de Janeiro, pp 33–39

Förstner U, Salomons W (1983) Heavy metals in contaminated sediments. Environ Technol Lett 1:1–9

Förstner U, Wittman GTW (1981) Metal Pollution in the Aquatic environment. Springer, Berlin Heidelberg New York

Fowler SW, Heyreaud M, La Rosa J (1978) Factors affecting methyl and inorganic mercury dynamics in mussels and shrimps. Mar Biol 46:267–276

Friberg L, Vostal J (eds) (1972) Mercury in the environment. CRC Press, Boca Raton

Fuge R, Pearce NIG, Perkins WT (1992) Mercury and gold mining. Nature 357:369

Fujika M (1963) Studies on the course that the causative agent of Minamata disease was formed, specially on the accumulation of mercury compound in the fish and shellfish of Minamata Bay. J Kumamoto Med Soc 37:494–521

Furch K, Junk WJ, Klinge H (1982) Unusual chemistry of natural waters from the Amazon region. Acta Cient Venezolana 33:269–273

Furutani A, Rudd JWM (1980) Measurement of mercury methylation in lake water and sediment samples. Appl Environ Microbiol 40:770–776

Furutani A, Rudd JWM, Turner MA (1980) Mercury methylation by fish intestinal contents. Appl Environ Microbiol 40:777–782

Galeano E (1981) As Veias Abertas da America Latina. Editora Paz e Terra, Rio de Janeiro

Gambrell RP (1994) Trace and toxic metals in wetlands – a review. J Environ Qual 23:883–891

Garrido I, Ribeiro GV, Costa IB, Azevedo J, Esteves MG, Amendola PC, Neves V (1989) Mineração: Uso do solo e meio ambiente na Amazônia. Proposta Metodol Rev Bras Geogr 51:25–51

GEDEBAM (1991) Mercury contamination in the Brazilian Amazon: background information on SOL3/GEDEBAM work in the Tapajós valley. Grupo de Estudos de Desenvolvimento da Amazônia. Belém, Pará

GESAMP (1986) Review of potentially harmfull substances: arsenic, mercury and selenium. Joint Group of Experts on Scientific Aspects of Marine Pollution

Gibbs RJ (1967) Amazon River: environmental factors that control its dissolved and suspended load. Science 156:1734–1737

Gibbs RJ (1973) Mechanism of metal transport in rivers. Science 180:274–280

Gibbs RJ (1976) Water chemistry of the Amazon River. Geochim Cosmochim Acta 36:1061–1066

Gill GA, Bruland KW (1990) Mercury speciation in surface freshwater systems in California and other areas. Environ Sci Technol 24:1392–1400

Gilmour CC, Henry E (1991) Mercury methylation in aquatic systems affected by acid deposition. Environ Pollut 71:131–169

Gilmour CC, Henry EA, Mitchell R (1992) Sulfate stimulation of mercury methylation in freshwater sediments. Environ Sci Technol 26:2281–2287

Glass GE, Sorensen JH, Schimidt KW, Rapp GR, Yap D, Fraser D (1991) Mercury deposition and sources for the upper Great lakes region. Water Air Soil Pollut 56:235–249

Golley FB, McGinnis JT, Clements RG, Child GI, Duever MJ (1975) Mineral cycling in a tropical forest ecosystem. Univ Georgia Press, Athens, p 248

Goulding M (1979) Ecologia da Pesca do Rio Madeira. CNPq/INPA. Manaus, AM, p 172

Guimarães JRD (1992) Padronização de Técnicas Radioquímicas Visando Estudos de Metilação e Volatilização do Hg em Sistemas Aquáticos de Áreas de Garimpo de Ouro na Região Amazônica. PhD Thesis, Federal University of Rio de Janeiro, p 109

Guimarães JRD, Malm O, Pfeiffer WC (1993a) Radiochemical measurements of net mercury methylation rates in sediments and soils near goldmining fields in the Tapajós river region, Brazilian Amazon. Proc Int Conf Heavy metals in the environment, Toronto 2:100–103

Guimarães JRD, Malm O, Pfeiffer WC (1993b) Radiochemical determination of net mercury methylation rates in sediment, water and soil samples from the Amazon Region. In: Abrão JJ, Wasserman JC, Silva-Filho EV (eds) Proc Int Symp Perspectives for environmental geochemistry in tropical countries, pp 413–416

Guimarães JRD, Malm O, Padovani C, Sanches MV, Forsberg BR, Pfeiffer WC (1994) A summary of data on net mercury methylation rates in sediment, water, soil and other samples from the Amazon region, obtained through radiochemical methods. In: Evironmental mercury pollution and its health effects in Amazon River Basin. Natl Inst Minamata Disease/Inst Biophysics, Univ Fed do Rio de Janeiro Rio de Janeiro, pp 94–99

Guimarães JRD, Malm O, Pfeiffer WC (1995) A simplified radiochemical technique for measurements of net mercury methylation rates in aquatic systems near gold mining areas, Amazon, Brazil. Sci Tot Environ 175:151–162

Gustin MS, Leonard GETT (1994) Mercury evasion from contaminated landscape elements, Carson River Superfund Site, Nevada. In: Proc Int Conf Mercury as a global pollutant, Whistler, BC, Sec III, Abstr

Hacon S (1991) Mercury contamination in Brazil, with emphasis on human exposure to mercury in the Amazon Region. FINEP, Rio de Janeiro, pp 91

Hacon S, Artaxo P, Gerab F, Yamasoe MA, Calixto RC, Conti LF (1995) Atmospheric mercury and trace elements in the region of Alta Floresta in the Amazon Basin. Water Air Soil Pollut 80:273–283

Hacon S, Lacerda LD, Pfeiffer WC, Carvalho D (eds) (1990) Riscos e Consequências do Uso do Mercúrio. FINEP/CNPq/MS/IBAMA, Rio de Janeiro, p 311

Hakanson L (1974) Mercury in some Swedish lake sediments. Ambio 3:37–43

Hall B, Schanger P, Lindqvist O (1991) Chemical reactions of mercury in combustion flue gases. Water Air Soil Pollut 56:3–14

Hannai M (1993) Mineração industrial, garimpo de ouro e meio ambiente no Brasil. In: Impactos Ambientais da Mineração e Metalurgia. CETEM/CNPq, Rio de Janeiro, pp 175–244

Haygrath PM, Jones KC (1992) Atmospheric deposition of metals to agricultural surfaces. In: Adriano DC (ed) Biogeochemistry of trace metals. Lewis, Boca Raton, pp 249–276

Hentshel T, Hruschka F, Priester M (1992) Mitigación de Emissioes de Mercúrio en la Pequeña Mineria Aurifera de Nariño, Colombia. CORPONARIÑo, Pasto, p 30

Hocking MB (1979) Uses and emissions of mercury in Canada. In: Effects of mercury in the Canadian environment. Natl Res Counc Canada, Ottawa, pp 50–75

Huckabee JW, Elwood JW, Hildebrand SG (1979) Accumulation of mercury by freshwater biota. In: Nriagu JO (ed) The biogeochemistry of mercury in the environment. Elsevier/North Holland Biomedical Press, Amsterdam, pp 277–307

Hutton M, Simon C (1986) The quantities of cadmium, lead, mercury and arsenic entering UK environment from human activities. Sci Tot Environ 57:129–150

Ikingura J (1994) Mercury contamination from gold mining in Tanzania. In: Evaluation on the role and distribution of mercury on ecosystems with special emphasis on tropical regions. SCOPE, Rio de Janeiro (unpubl)

INPE (1992) Deforestation rates in the Amazon. Inst Nac Pesq Espaciais, São José dos Campos

Inskip MJ, Piotrowski JK (1985) Review of the health effects of methyl-mercury. J Appl Toxicol 5:113–133

Iverfeldt A (1991) Occurrence and turnover of atmospheric marcury over the Nordic countries. Water Air Soil Pollut 56:252–265

Jackson TA (1986) Methylmercury levels in a polluted prairie river-lake system: seasonal and site-specific variations and the dominant influence of trophic conditions. Can J Fish Aquat Sci 43:1873–1887

Jackson TA (1988) The mercury problem in recently formed reservoirs of northern Manitoba (Canada): effects of impoundment and other factors on the production of methylmercury by microorganisms in sediments. Can J Fish Aquat Sci 45: 97–121

James LP (1994) The mercury tromol mill: an innovative gold recovery technique and a possible environmental concer. J Geochem Explor 50:493–500

Jensen S, Jernelov A (1969) Biological methylation of mercury in aquatic organisms. Nature 223:753–754

Jernelov A, Hansson C, Linse L (1976) Mercury in fish in Varmland. An investigation of the effect of pH and the total phosphorus on the measured variation. IVL Rep B 282. Swedish Institute for Water and Air Pollution Research, Stockholm, p 7

Jernelov A, LannG, Wennergren T, Fagerstrom B, Andersson R (1972) Analysis of methyl mercury concentrations in sediments from the St. Clair system. Unpubl Report Swed Water Air Poll Res Inst, Stockholm

Johnels A, Tyler G, Wertermark T (1979) A history of mercury in Swedish fauna. Ambio 8:160–168

Jordan CF, Golley FB, Hall J, Hall J (1980) Nutrient scavenging of rainfall by the canopy of an Amazonian rain forest. Biotropica 12:61–66

Junk WJ, Furch K (1980) Química da água e macrofitas aquáticas de rios e igarapés na Bacia Amazônica e nas areas adjacentes. Parte I. Acta Amazônica 10:611–633

Kauffman YJ, Setzer A, Ward D, Tanre B, Holbe N, Menzrl P, Pereira MC, Rasmussen R (1992) Biomass burning airborne and spaceborne experiment in the Amazonas (BASE-A). J Geophy Res 97D13:14581

Kelly M (1988) Mining and the freshwater environment. Elsevier, London, p 231

Kersten M (1988) Geochemistry of priority pollutants in anoxic sludge. In: Salomons W, Forstner (eds) Chemistry and biology of solid wastes. Springer, Berlin Heidelberg New York, pp 170–213

Kim K-H, Lindberg S, Hanson PJ, Meyers TP, Owens J (1993) Application of micrometeorological methods to measurements of mercury emissions over contaminated soils. Proc Int Conf Heavy metals in the environment, Toronto 1:328–331

Korthals ET, Winfrey MR (1987) Seasonal and spatial variation in mercury methylation and demethylation in an oligotrophic lake. Appl Environ Microbiol 53:2397–2404

Kothny E (1974) The three-phase equilibrium of mercury in nature. Adv Chem Ser 123:48–91

Kozlowiski TT (1984) Plate responses to flooding of soil. BioScience 34:162–167

Laborão J (1990) Importação, comercialização e controle de mercúrio no País. In: Hacon S, Lacerda LD, Pfeiffer WC, Carvalho D (eds) Riscos e Consequências do Uso do Mercúrio. FINEP/MS/CNPq/IBAMA. Rio de Janeiro, pp 141–144

Lacerda LD (1990) Ciclo biogeoquímico do mercúrio na Amazônia. In: Hacon S, Lacerda LD, Pfeiffer WC, Carvalho D (eds) Riscos e Consequências do Uso do Mercúrio. UFRJ. Rio de Janeiro, pp 80–90
Lacerda LD (1995) Amazon mercury emissions. Nature 374:21–22
Lacerda LD (1995) Evolution of mercury contamination in Brazil. Water Air Soil Pollut (in press)
Lacerda LD (1995) Global emissions of Hg from gold and silver mining. Water Air Soil Pollut (in press)
Lacerda LD, Pfeiffer WC, Silveira EG, Bastos WR, Souza CMM (1987) Contaminação por mercúrio na Amazônia: analise preliminar do Rio Madeira, RO. An I Congr Bras Geoquim 2:295–299
Lacerda LD, Pfeiffer WC, Ott AT, Silveira EG (1989) Mercury contamination in the Madeira River, Amazon: mercury inputs to the environment. Biotropica 21:91–93
Lacerda LD, DePaula FC, Ovalle ARC, Pfeiffer WC, Malm O (1990) Trace metals in fluvial sediments of the Madeira River watershed, Amazon, Brazil. Sci Tot Environ 97/98:525–530
Lacerda LD, Salomons W (1991) Mercury in the Amazon. A chemical time bomb? Dutch Ministry of Housing, Physical Planning and the Environment, The Hague, The Netherlands
Lacerda LD, Marins RV, Souza CMM, Rodrigues S, Pfeiffer WC, Bastos WR (1991a) Mercury dispersal in water, sediments and aquatic biota of a gold mining tailings drainage in Pocone, Brazil. Water Air Soil Pollut 55:283–294
Lacerda LD, Salomons W, Pfeiffer WC, Bastos WR (1991b) Mercury distribution in sediment profiles of remote high Pantanal lakes, central Brasil. Biogeochemistry 14:71–77
Lacerda LD, Pfeiffer WC, Bastos WR (1991c) Mercury dispersal in the Pocone Region, Mato Grosso State, central Brazil. Ciência e Cultura 43:317–320
Lacerda LD, Carvalho CEV, Rezende CE, Pfeiffer WC (1993) Mercury in sediments of the Paraíba do Sul River estuary and continental shelf, SE Brazil. Mar Poll Bull 26:220–222
Lacerda LD, Bidone ED (1993) Seasonal variation in iron, manganese and copper in the Itacaiúnas-Parauapebas Rivers, Carajás region, Amazon. In: Proc 9th Int Conf Heavy metals in the environment, Toronto 1:173–176
Lacerda LD, Bidone ED, Guimarães AF, Pfeiffer WC (1994) Mercury distribution in fish from the Itacaiúnas-Parauapebas River system, Carajás region, Amazon. An Acad Brasil Ciênc 66:373–379
Lacerda LD, Prieto G, Marins RV, Azevedo SLN, Pereira MC (1995a) Anthropogenic mercury emissions to the atmosphere in Brazil. In: Proc 10th Int Conf Heavy metals in the environment, Hamburg (in press)
Lacerda LD, Malm O, Guimarães JRD, Salomons W (1995b) Mercury and the new gold rush in the South In: Salomons W, Stigliani W (eds) Biogeodynamics of pollutants. Springer, Berlin Heidelberg (in press)
Landner L (1971) Biochemical models of the biological methylation of mercury as suggested from methylation studies in-vivo in *Neurospora crassa*. Nature 230:452–453
Lane PA, Crowell MJ, Graves MC (1988) Heavy metal removal from gold mining and tailings effluents using indigenous aquatic macrophytes. (Phase I). CNMET Spec Publ SP88-23, pp 3–37
Lechler PJ, Miller JR (1993) The dispersion of mercury, gold, and silver from contaminated mill tailings at the Carson River mercury superfund site, west-central Nevada, USA. In: Abrão JJ, Wasserman JC, Silva-Filho EV (eds) Proc Int Symp Perspectives for environmental geochemistry in tropical countries, pp 433–436

Lechler PJ (1993) Mercury vapor sampling at the Carson River superfund site. Int Conf Heavy metals in the environment, Toronto 1:377–380

Lee YH, Hultberg H (1990) Methylmercury in some swedish surface waters. Environ Toxicol Chem 9:833–841

Lee YH, Hultberg H, Andersson I (1985) Catalytic effect of various metal ions on the methylation of mercury in the presence of humic substances. Water Air Soil Pollut 25:391–400

Lesenfants Y (1994a) Informations about the mercurial pollution in Venezuelan Guayana due to gold mining. Bioma (Fund Venezolana Conserv Diversidad Biol), Caracas, p 3

Lesenfants Y (1994b) Gold mining and mercury contamination in Veezuela. In: Environmental mercury pollution and its health effects in Amazon River Basin. Natl Inst Minamata Disease and Inst Biophysics of the Univ Federal do Rio de Janeiro Rio de Janeiro, pp 17–20

Lima ACR (1990) Riscos e consequências do uso do mercúrio: a situação do Rio de Janeiro. In: Hacon S, Lacerda LD, Pfeiffer WC, Carvalho D (eds) Riscos e Consequências do Uso do Mercúrio. FINEP/MS/CNPq/IBAMA. Rio de Janeiro, pp 268–272

Lindberg SE, Harris RC (1974) Mercury-organicmatter associations in estuarine sediments and interstitial water. Environ Sci Technol 8:459–462

Lindqvist O, Rhode H (1984) Atmospheric mercury – a review. Tellus 37B:136–159

Lindqvist O, Johanso K, Aastrup M, Andersson A, Bingmark L, Hovsenius G, Hakanson L, Iverfeldt A, Meili M, Timm N (1991) Mercury in the Swedish environment. Recent research on causes, consequences and corrective methods. Water Air Soil Pollut 55:143–177

Lindqvist O, Jernelov A, Johanson K, Rodhe H (1984) Mercury in the Swedish environment. Global and local sources. Rep PM 1816, Natl Swed Environ Protect Board, Solna, Sweden

Lock CGW (1901) Gold milling principles and practice. E & FN Spon, London, p 632

Lodenius M (1985) The mercury problem and fishing in Finland. In: Hall DO, Myers N, Margaris NS (eds) Economics of ecosystem management. Junk, Dordrecht, pp 99–103

Lodenius M (1991) Mercury concentrations in an aquatic ecosystem during twenty years following abatement of the pollution source. Water Air Soil Pollut 56:323–332

Lodenius M, Seppanen A, Herranen M (1983) Accumulation of mercury in fish and man from reservoirs in northern Finland. Water Air Soil Pollut 19:237–246

Madsen PP (1981) Peat bog records of atmospheric mercury deposition. Nature 293:127–130

Mallas J, Benedicto N (1986) Mercury and gold mining in the Brazilian Amazon. Ambio 15:248–249

Malm O (1991) Contaminação ambiental e humana por mercúrio na região garimpeira do Rio Madeira, Amazônia. PhD Thesis, Inst Biofísica, Univ Fed Rio de Janeiro, Rio de Janeiro

Malm O (1993) In: Mathis A, Dhaag R (eds) Consequências da Garimpagem no Âmbito Social e Ambiental da Amazônia. Buntstiff e.v., Katalise, FASE, Belém, pp 21–34

Malm O, Pfeiffer WC, Souza CMM, Reuther R (1990) Mercury pollution due to gold mining in the Madeira River Basin, Brazil. Ambio 19:11–15

Malm O, Pfeiffer WC, Souza CMM (1991) Main pathways of mercury in the Madeira River area, Rondonia, Brazil. Proc Int Conf Heavy metals in the environment, Edinburgh 1:515–518

Malm O, Branches FJP, Castro MB, Pfeiffer WC (1992) Mercury contamination in riverine population through ingestion of fish in the Tapajós Basin, Amazon, Brazil. Abstr 4th Annu Meet Int Soc Environment epidemiology, México Session: 03.1

Malm O, Guimarães JRD, Pfeiffer WC (1993) Accumulation of metallic mercury and natural amalgams findings in Madeira River basin bottom sediments, Amazon. In: Abrão JJ, Wasserman JC, Silva-Filho EV (eds) Proc Int Symp Perspectives for environmental geochemistry in tropical countries, pp 391–393

Malm O, Castro MB, Bastos WR, Branches FJP, Guimarães JRD, Zuffo CE, Pfeiffer WC (1995a) An assessment of Hg pollution in different gold mining areas, Amazon, Brazil. Sci Tot Environ 175: 141–150

Malm O, Branches FJP, Akagi H, Castro MB, Pfeiffer WC, Harada M, Bastos WR, Kato H (1995b) Mercury and methylmercury in fish and human hair from Tapajós River Basin, Brazil. Sci Tot Environ 175: 127–140

Mannio J, Verta M, Kortelainen P, Rekolainen S (1986) The effect of water quality on the mercury concentration of northern pike (*Esox lucius* L.) in Finnish forest lakes and reservoirs. Publ Water Res Inst Finland 65: 32–43

MARC (1987a) Biological monitoring. MARC Rep no 32. Monitoring and Assessment Research Centre, London

MARC (1987b) Biological monitoring. Animals. MARC Rep no 37. Monitoring and Assessment Research Centre, London

Marins RV, Tonietto GB (1995) An evaluation of the sampling and measurements techniques of mercury in the air. 5th Brazilian Geochemical Congr, Niteroi, pp 17–20

Marins RV, Imbassay JA, Pfeiffer WC, Bastos WR (1991) Contaminação atmosférica de mercúrio em area produtora de ouro no distrito de Poconé, Mato Grosso, MT. Proc I Int Simp Environ Studies Tropical Forest, Manaus 1: 199–203

Marins RV, Imbassahy JA, Pfeiffer WC, Bastos WR (1990) Preliminary study on mercury contamination in the urban atmosphere of a gold producing area in Poconé, Mato Grosso (MT). In: 1st Int Symp Environ Stud Tropical humid forests, Manaus, p 6

Martinelli LA, Ferreira JR, Forsberg BR, Victoria RL (1988) Mercury contamination in the Amazon: a gold rush consequence. Ambio 17: 252–254

Martinelli LA, Victoria RL, Mortati J, Forsberg BR, Bonassi JA, Oliveira F, Tancredi AC (1988a) Nutrient fluxes in some Rondonia Rivers, Madeira Basin. Acta Limnol Bras 2: 761–773

Martins AF, Zanella R (1990) Estudo analítico-ambiental na regiao carboenergética de Candiota, Bagé (RS). Cienc Cult 42: 264–270

Mason RP, Fitzgerald WF, Morel FMM (1994) The biogeochemical cycling of elemental mercury. Anthropogenic influences. Geochim Cosmochim Acta 58: 3191–3198

Mason RR, Morel FMM (1993) An assessment of the principal pathways for oxidation of elemental mercury and the production of methyl-mercury in Brazilian waters affected by goldmining activities. In: Abrão JJ, Wasserman JC, Silva-Filho EV (eds) Proc Int Symp Perspectives for environmental geochemistry in tropical countries, pp 413–416

Matsumura F, Gotoh Y, Boush GM (1971) Phenylmercuric acetate: metabolic conversion by microorganisms. Science 173: 49–51

Meili M (1991a) Mercury in boreal forest lake ecosystems. PhD Thesis, Acta Univ Ups, Comprehensive Summaries of Uppsala Diss, Faculty of Science 336 Uppsala University, Sweden, p 36

Meili M (1991b) The coupling of mercury and organic matter in the biogeochemical cycle: towards a mechanistic model for boreal forest zone. Water Air Soil Pollut 56:333–347

Meili M (1991c) Fluxes, pools and turnover of mercury in Swedish forest lakes. Water Air Soil Pollut 56:119–127

Meili M (1991d) In situ assessment of trophic levels and transfer rates in aquatic food webbs using chronic (Hg) and pulsed (Chernobyl ^{137}Cs) contaminants. Verh Int Verein Limnol 24:2970–2975

Meili M, Iverfeldt A, Hakanson A (1991) Mercury in the surface water of Swedish forest lakes – concentrations, speciation and controlling factors. Water Air Soil Pollut 56:439–453

Mellor JW (1952) A comprehensive treatise on inorganic and theoretical chemistry, vol IV. Longmans, Green, Co, London

Mendoza V (1990) Geología ambiental y el desarollo de recursos minerales. Bol Geominas 20:27–60

Miller JR, Lechler PJ, Rowland J (1993a) Heavy metal transport by physical processes in the Carson River Valley, west-central Nevada, USA: implications to the distribution and storage of metal pollutants in tropical environments. In: Abrão JJ, Wasserman JC, Silva-Filho EV (eds) Proc Int Symp Perspectives for environmental geochemistry in tropical countries, pp 131–136

Miller JR, Lechler PJ, Desileta M, Rowland J, Hsu LC, Price JG (1993b) Quantity and distribution of mining related trace metals in Lohantan Reservoir, west central Nevada. Proc Int Conf Heavy metals in the environment, Toronto 2:251–254

Ming G (1994) In: Evaluation on the role and distribution of mercury on ecosystems with special emphasis on tropical regions. SCOPE, Rio de Janeiro, (unpubl)

Mitra S (1986) Mercury in the ecosystem. Transtech Publ, Columbus, Ohio, p 327

MME (1992) Perfil Energético Brasileiro. Ministério das Minas e Energia, Brasília, DF

Mohlenberg F, Riisgard HU (1988) Partitioning of inorganic and organic mercury in cockles *Cardium edule* (L.) and *C. glaucum* (Bruguiere) from a chronically polluted area: influence of size and age. Environ Pollut 55:137–148

Moore JW, Sutherland DJ (1980) Mercury concentrations in fish inhabiting two polluted lakes in northern Canada. Wat Res 14:903–907

Mora SJ, Patterson JE, Bibby DM (1993) Baseline atmospheric mercury studies at Ross Island, Antartica. Antartic Sci 5:323–326

Moriarty F (1974) Pollutants and animals. A factual perspective. George Allen and Unwin, London

Mudroch A, Clair TA (1986) Transport of arsenic and mercury from gold mining activities through an aquatic system. Sci Total Environ 57:205–216

Mudroch A (1983) Distribution of major elements ad metals in sediment cores from the western basin of Lake Ontario. J Great Lakes Res 9:125–133

Müller G (1993) Long distance mercury transport in Rio Tapajós, Pará, Brazil. In: Abrão JJ, Wasserman JC, Silva-Filho EV (eds) Proc Int Symp Perspectives for environmental geochemistry in tropical countries, pp 363–364

Nadkarni NH (1984) Epiphyte biomass and nutrient capita of a neotropical elfin forest. Biotropica 16:249–256

Nelson H, Larsen BR, Jenne EA, Sorg DH (1977) Mercury dispersal from Lode sources in the Kuskokwim River drainage, Alaska. Science 198:820–824

Nico LG, Taphorn DC (1994) Mercury in fish from gold-mining regions in upper Cuyuni River system, Venezuela. Fresenius Environ Bull 3:287–292

Nilsson B, Gerhardsson L, Nordberg GF (1990) Urine mercury levels and associated symptoms in dental personnel. Sci Tot Environ 94:179–185
Nriagu JO, Pacyna JM (1988) Quantitative assessment of worldwide contamination of air, water and soils by trace metals. Nature 333:134–139
Nriagu JO (1979) Global inventory of natural and anthropogenic emissions of trace metals to the atmosphere. Nature 279:409–411
Nriagu JO (1989) A global assessment of natural sources of atmospheric trace metals. Nature 338:47–49
Nriagu JO (1990) Global metal pollution poisoning the biosphere. Environment 32:7–33
Nriagu JO (1993a) Mercury pollution from silver mining in colonial South America. In: Abrão JJ, Wasserman JC, Silva-Filho EV (eds) Proc Int Symp Perspectives for environmental geochemistry in tropical countries, pp 365–368
Nriagu JO (1993b) Legacy of mercury pollution. Nature 363:589
Nriagu JO (1994) Mercury pollution from past mining of silver and gold in the Americas. Sci Tot Environ 149:167–181
Nriagu JO, Pfeiffer WC, Malm O, Souza CMM, Mierle G (1992) Mercury pollution in Brazil. Nature 356:389
Nuorteva P, Lodenius M, Nuorteva SL (1979) Decrease in mercury levels of *Esox lucius* (L.) and *Abramis farenus* (L.) (teleostei) in the Haneenkyro water course after the phenylmercury ban in Finland. Aquilo Ser Zool 19:97–100
Ogilvie DM (1981) Mercury clean-up in Canada's Lake St. Clair: only a partial success. Ambio 10:350–351
Otani Y, Kanaoka C, Usui C, Matsui S, Emi H (1986) Adsorption of mercury vapour on particles. Environ Sci Technol 20:735–738
Pacyna JM, Münch J (1991) Anthropogenic mercury emission in Europe. Water Air Soil Pollut 56:51–61
Pacyna JM (1984) Atmospheric trace elements from natural and anthropogenic sources. In: Nriagu JO, Davidson CI (eds) Toxic metals in the atmosphere. Wiley, New York
Padberg S (1990) Mercury determinations in samples from Tapajó s (Itaituba). Inst Angewandte Physikalische Chemie, Julich, FRG 13 pp
Paterson C (1971) Metal stocks. Am Antiq
Payne AI (1986) The ecology of tropical lakes and rivers. Wiley, Chichester, pp 301
Pelletier E (1985) Mercury-selenium interactions in organisms: a review. Mar Environ Res 18:111–132
Pestana MHD, Formoso MLL, Teixeira EC (1993) Studies of heavy metals in gold and copper mining areas of Camaquã River Basin, southern Brazil. In: Abrão JJ, Wasserman JC, Silva-Filho EV (eds) Proc Int Symp Perspectives for environmental geochemistry in tropical countries, pp 339–341
Petersen G, Eppel D, Grassl H, Iverfeldt A, Misra PK, Bloxmam R, Wong S, Schroeder WH, Voldner E, Pacyna JM (1989) Model studies on the atmospheric transport and deposition of mercury. 7th Int Conf Heavy metals in the environment 1: 48–52
Petrick FR (1993) Bindung und akkumulation von quecksilber in den vom goldabbau kontaminierten flussedimenten des rio Madeira, Rondônia, Brasilien. PhD Thesis, Ludwig Maximilians Universität, München, p 112
Pfeiffer WC, Lacerda LD (1988) Mercury inputs into the Amazon Region, Brazil. Environ Technol Lett 9:325–330

Pfeiffer WC, Lacerda LD, Malm O, Souza CMM, Silveira EG, Bastos WR (1989b) Mercury concentrations in inland waters of Rondônia, Amazon, Brazil. Sci Tot Environ 87/88: 233–240

Pfeiffer WC, Lacerda LD, Malm O, Souza CMM, Silveira EG, Bastos WR (1991) Mercury in the Madeira River ecosystem, Rondonia, Brazil. For Ecol Manage J 38: 239–245

Pfeiffer WC, Malm O, Souza CMM, Bastos WR, Torres JP (1989a) Mercury contamination in goldminning areas of Rio de Janeiro State, Brazil. Proc Int Conf Heavy metals in the environment, Geneve 1: 222–225

Pfeiffer WC, Petrick FR, Malm O, Guimarães JRD (1993) Biogeochemistry of mercury in bottom sediments from the Madeira River Basin, Rondônia. In: Abrão JJ, Wasserman JC, Silva-Filho EV (eds) Proc Int Symp Perspectives for environmental geochemistry in tropical countries, pp 425–427

Pinto JA (1990) Impactos socio-ecológicos da mineração e da garimpagem na Amazônia Oriental (Estado do Pará). In: Flores CM, Mitschein TA (orgs) Realidades Amazônicas no Fim do Século XX. Ser Cooperação Amazônica no 5 UNAMAZ, Belem, pp 434–459

Priester M (1993) Tecnologia de meio ambiente para mini-mineração e sua difusão. A experiência de um projeto GTZ para evitar emissões de mercúrio na mineração primária de ouro no sul da Colombia. In: Mathis A, Rehaag R (orgs) Conseqüências da Garimpagem no Ambito Social e Ambiental da Amazônia. FASE/buntstft e. v./KATALISE, Belém pp 102 –112

Prieto GR (1995) PhD Diss, Univ Belém, Brazil

Prokopovich NP (1984) Occurrence of mercury in dredge tailings near Falosm South Canal, California. Bull Assoc Eng Geol 21: 531–543

Quevauviller P, Donard OFX, Wasserman JC, Martin FM, Schneider J (1992) Occurrence of methylated tin and dimethyl mercury compounds in a mangrove core from Sepetiba Bay, Brazil. Appl Organometallic Chem 6: 221–228

Ramlal PS, Rudd JMW, Furutani A, Xun L (1985) The effect of pH on methylmercury production and decomposition in lake sediment. Can J Fish Aquat Sci 42: 685–692

Ramlal PS, Rudd JWM, Hecky RE (1986) Methods for measuring specific rates of mercury methylation and degradation and their uses in determining factors controlling net rates of mercury methylation. Appl Environ Microbiol 51: 110–114

Ramos JFF, Costa MQ (1990) Distribuição de mercúrio em dois garimpos do Estado do Pará. In: Hacon S, Lacerda LD, Pfeiffer WC, Carvalho D (eds) Riscos e Conseqüências do uso do Mercúrio. UFRJ Rio de Janeiro, pp 70–79

Rasmussen PE (1994) Current methods of estimating atmospheric mercury fluxes in remote areas. Environ Sci Technol 28: 2233–2241

Rekolainen S, Verta M, Liehu A (1986) The effect of airborne mercury contents in some Finnish forest lakes. Publ Water Res Inst Finland 65: 11–20

Renzoni A (1987) Mercury levels in human hair and their relevance to health. Proc Int Conf Heavy metals in the environment, New Orleans 2: 80–82

Robinson JB, Tuovinen OH (1984) Mechanisms of microbial resistance and detoxification of mercury and organomercury compounds: physiological, biochemical and genetic analyses. Microbiol Rev 48: 95–124

Rodrigues BA, Lenzi E, Luchese EB, Rauber T (1992) Níveis de concentração de mercúrio em águas do Rio Paraná/Baía, Região de Porto Rico. Acta Limnol Bras 4: 255–260

Rodrigues RM, Mascarenhas AF, Ichihara AH, Souza TMC, Bidone ED, Bellia V, Hacon S, Silva ARB, Braga JB, Stilianidi B (1991) Estudo dos impactos ambientais decorrentes do extrativismo mineral e poluição mercurial no Tapajós: Pré-diagnóstico. CETEM/CNPq, Rio de janeiro. Ser Tecnol Ambiental 4, pp 1–218

Rodrigues S, Maddock JEL (1993) A survey of heavy metal distribution in sediments and water from the gold prospecting region of Poconé (MT), Brazil. In: Abrão JJ, Wasserman JC, Silva-Filho EV (eds) Proc Int Symp Perspectives for environmental geochemistry in tropical countries, pp 461–466
Rodrigues S (1994) Determinação dos níveis de "background" e avaliação do grau de contaminação por metais pesados em sub-bacias hidrográficas das regiões garimpeiras de Poconá (MT) (Hg, Cu, Pb, Zn, Fe e Mn) e Alta Floresta (Hg) MSc Thesis Univerisdade Federal fluminense, Niterói, p 80
Rogers RD (1976) Methylation of mercury in agricultural soils. J Environ Qual 5:454–458
Rogers RD (1977) Abiological methylation of mercury in soil. J Environ Qual 6:463–467
Rose TK (1915) The metallurgy of Gold. C Griffin, New York
Roulet M, Lucotte M (1994) Biogeochemistry of mercury in inundated equatorial forest, French Guiana. In: Proc Int Conf Mercury as a global pollutant, Whistler BC, Sec 12A Abstr
Salomons W, Förstner U (1984) Metals in the hydrocycle. Springer, Berlin Heidelberg New York, p 349
Salomons W, Stigliani WM (eds) (1995) Biogeodynamics of pollutants in soils and sediments. Springer, Berlin Heidelberg New York, pp 320
Schoroeder WH (1982) Sampling and analysis of mercury and its compounds in the atmosphere. Environ Sci Technol 16:394–400
Schoroeder WH, Lindqvist O, Munthe J, Xiao Z (1992) Volatilization of mercury from lake surfaces. Sci Tot Environ 125:47–66
SCOPE (1985) Rapporteur's report – mercury. Sci Comm Problems Environ, Toronto, p 16
Scrudato RJ, Long D, Weinbloom R (1987) Mercury contribution to an Adirondack lake. Environ Geol Water Sci 9:131–137
SEMA (1988) Avaliação da Degradação Ambiental nas Áreas de Extração de Ouro no Estado do Pará. Projeto Mercúrio. Secretaria Especial do Meio Ambiente, Brasília DF, p 180
Shrestha KP, Quilarque XR (1989) A preliminary study of mercury contamination in the surface soil and river sediment of the Roscio District, Bolivar State, Venezuela. Sci Tot Environ 79:233–239
Siegel SM (1973) Metal ions in biological systems. Marcel Decker, New York
Siegel SM, Siegel BZ, Puerner N, Speitel T, Thorarinsson F (1975) Water and soil biotic mercury distribution. Water Air Soil Pollut 4:9–17
Siegel SM, Siegel BZ, Lipp C, Kruckberg A, Towers GHN, Warre H (1985) Indicator plant-soil mercury pettern in a mercury rich mining area in British Columbia. Water Air Soil Pollut 25:73–78
Siegel SM, Siegel BZ, Barghigiani C, Aratani K, Penny P, Penny D (1987) A contribution of the environmental biology of mercury accumulation in plants. Water Air Soil Pollut 33:65–72
Silva AS, Oliveira AOF, Goncalves HS, Anjos JR, Amaral Estrela V, Barbosa AC, Ponce GAE, Ferreira GA, Dorea JG (1990) Avaliação da poluição por mercúrio nos garimpos de Paracatu, MG. In: Hacon S, Lacerda LD, Pfeiffer WC, Carvalho D (eds) Riscos e Consequências do Uso do Mercúrio. FINEP/CNPq/MS/IBAMA, Rio de Janeiro, pp 30–45
Simola H, Lodenius M (1982) Recent increase in mercury sedimentation in a forest lake attributable to peatland drainage. Bull Environ Contam Toxicol 29:298–305
Simonsen RC (1962) Historia Econômica do Brasil. Companía Editora Nacional, São Paulo

Sioli H (1950) Das Wasser in Amazonasgebeit. Forsch Fortschr 26:274–280
Slemr F, Seiler W, Shuster G (1991) Latitudunal distribution of mercury over the Atlantic Ocean. J Geophys Res 86:1159–1162
Slemr F, Langer E (1992) Increase in global atmospheric concentrations of mercury inferred from measurements over the Atlantic Ocean. Nature 365:434–437
Smith RB (1869) The gold fields and mineral districts of Victoria. Facsimile Reprint, 1980. Queensberry Hall Press, Melbourne
Smith RG, Vorwald AB, Patil LS, Moorney TF 1970 Effects of exposure to mercury in the manufacture of chlorine. Am Ind Hyg Assoc J 31:687–700
Souza VP, Lins FAF (1989) Recuperação de Ouro por Amalgamação e Cianetação. CETEM/CNPq, Ser Tec no 44, Rio de Janeiro, p 27
Stalard RF, Edmond JM (1983) Geochemistry of the Amazon. The influence of geology and weathering environments on dissolved load. J Geophys Res 88:9671–9688
Stallard RF (1985) River chemistry, geology, geomorphology and soils in the Amazon and Orinoco basins. In: JI Drever (ed) Chemistry of weathering. D Heildel, Hingham, Massachusetts, pp 239–316
Steffan RJ, Korthals ET, Winfrey MR (1988) Effects of acidification on mercury methylation, demethylation and volatilization in sediment of an acid-susceptible lake. Appl Environ Microbiol 54:2003–2009
Steines E, Andersson EM (1991) Atmospheric deposition of mercury in Norway: Temporal and spatial trends. Water Air Soil Pollut 56:391–404
Stevenson FJ (1977) Nature of divalent transition metal complexes of humic acids as revealed by a modified potentiometric tritation method. Soil Sci 123:10–17
Sukhenko SA, Papina TS, Pozdnjakov SR (1992) Transport of mercury by the Katun River, West Siberia. Hydrobiologia 228:23–28
Tayrayev TT (1991) man-made dispersion train of gold and mercury in Golet-Taiga terrain. Dokl Akad Auk SSR 317:719–722
Thibodeaux LJ (1979) Chemodynamics – environmental movement of chemicals in air, water and soil. Wiley, New York, pp 491
Thornton I, Cleary D, Kazantis G (1994) Mercury contamination from gold mining in the Brazilian Amazon: Health implications. In: Varnavas SP (ed) 6th Int Conf Environ Contam, Delphi, pp 26–33
Thornton I, Cleary D, Worthington S (1992) Mercury contamination in the Brazilian Amazon. A cooperative research study conducted by GEDEBAM, Brazil e Sol 3 (Switzerland and Luxemburg), p 45
Torres EB (1992) Environmental and health survelliance of mercury use in small scale gold processing industries in the Philippines. In: Proc Int Symp Epidemiol Stud Environ Pollut and health effects of methylmercury, Natl Inst for Minamata Disease, Kumamoto pp 56–65
Torres EB (1994) Epidemiological investigation of mercury exposure and health effects in the Philippines. In: Proc Int Worksh on Environmental mercury pollution and its health effects in Amazon River Basin, Natl Inst Minamate Disease, Inst Biofisica Univ Fed Rio de Janeiro, Rio de Janeiro, pp 3–9
Tumpling WV, Wilken RD, Einax J (1993a) Pathways and contaminations of mercury in the Pantanal area, Brazil. In: Abrão JJ, Wasserman JC, Silva-Filho EV (eds) Proc Int Symp Perspectives for environmental geochemistry in tropical countries, pp 403–406
Tümpling WV, Wilken RD, Einax J (1993b) Mercury contamination in the Pantanal area, Brazil. Proc Int Conf Heavy metals in the environment, Toronto 2:74–77

Tümpling WV, Wilken RD, Einax J (1995) Mercury contamination in the northern Pantanal region, Mato Grosso, Brazil. J Geochem Explor 52:127–134
USGS (1968) Minerals yearbook. United States Geological Survey, Washington, DC
USGS (1970) Mercury in the environment. United States Geological Survey, Washington, DC
Vandal GM, Fitzgerald WF, Boutron CF, Candelone JP (1993) Variations uin mercury deposition to Antartica over the past 34,000 years. Nature 362:621–623
Veiga MM, Fernandes FRC (1991) (orgs) Poconé: Um campo de estudos do impacto ambiental do garimpo. Centro de Tecnologia Mineral (CETEM-CNPq). Rio de Janeiro, p 113
Veiga MM, Meech JA, Onate N (1994) Mercury pollution from deforestation. Nature 368:816–817
Walter CM, June FC, Brown HG (1973) Mercury in fish, sediments and water in Lake Oahe, South Dakota. JWPCF 45:2203–2210
Weber JH, Reisinger H, Stoeppler M (1985) Methylation of mercury(II) by fulvic acid. Environ Technol Lett 6:203–208
Welcomme RL (1979) Fisheries ecology of floodplain rivers. Longman, London, p 317
WHO (1990) IPCS – methylmercury. Environmental Health Criteria, World Health Organization 101:42–58
WHO (1991) IPCS – inorganic mercury. Environmental Health Criteria, World Health Organization 118:84–114
WHO (1976) Environmental health criteria no 1, Mercury. World Health Organization, Geneva, p 131
WHO (1980) Recomended health based limits in occupational exposure to heavy metals. Tech Rep Ser No 647, Geneva
WHO (1991) IPCS – inorganic mercury. Environmental Health Criteria 118, World Health Organization, Geneva, pp 168
Winfrey MR, Rudd JWM (1990) Environmental factors affecting the formation of methylmercury in low pH lakes. Environ Toxicol Chem 9:853–869
Wise EM (1966) Gold and gold compounds. In: Kirk-Ohmer Encyclopedia of chemical technology, 2nd edn. Wiley, New York, pp 681–694
Xun L, Campbell NER, Rudd JWM (1987) Measurement of specific rates of net methylmercury production in the water column and surface sediments of acidified and circumneutral lakes. Can J Fish Aquat Sci 44:750–757
Yshuan YS (1994) In: Evaluation on the role and distribution of mercury on ecosystems with special emphasis on tropical regions. SCOPE, Rio de Janeiro (unpubl)
Zapata JQ (1994) Environmental impacts study of gold mining in Nueva Esperanza (ARARAS) Departamento de Pando, in the Madeira River Bolivian-Brazilian border. In: Evironmental mercury pollution and its health effects in Amazon River Basin. Natl Inst Minamata Disease and Inst Biophysics of the Univ Federal do Rio de Janeiro Rio de Janeiro, pp 23–24

Subject Index

Springer and the environment

At Springer we firmly believe that an international science publisher has a special obligation to the environment, and our corporate policies consistently reflect this conviction.
We also expect our business partners – paper mills, printers, packaging manufacturers, etc. – to commit themselves to using materials and production processes that do not harm the environment. The paper in this book is made from low- or no-chlorine pulp and is acid free, in conformance with international standards for paper permanency.

Printing: Saladruck, Berlin
Binding: Buchbinderei Lüderitz & Bauer, Berlin